T0211413

Rock Mass Response to Mining Activities

Geomechanics Research
Series Editor: Tsuyoshi Ishida
Department of Civil and Earth Resources Engineering, Kyoto University, Kyoto, Japan

ISSN print : 0929–4856
ISSN online: 2154–5782

Rock Mass Response to Mining Activities

Inferring Large-Scale Rock Mass Failure

Tadeusz Szwedzicki

Independent Consultant in Mining Geomechanics,
Sorrento, Australia

CRC Press
Taylor & Francis Group
Boca Raton London New York

CRC Press is an imprint of the
Taylor & Francis Group, an **informa** business

A BALKEMA BOOK

Cover illustrations:

Photo top left: Damage to the hangingwall at the shoulder of a squat pillar.
Photo top right: Pillar punching the roof.
Photo bottom left: Sinkhole, Warrego mine (1989).
Photo bottom right: Asymmetrical closure of a crosscut due to approaching caving front.

All photos by T. Szwedzicki

Published by:
CRC Press/Balkema
P.O. Box 447, 2300 AK Leiden, The Netherlands
e-mail: Pub.NL@taylorandfrancis.com
www.crcpress.com – www.taylorandfrancis.com

First issued in paperback 2021

© 2018 by Taylor & Francis Group, LLC
CRC Press/Balkema is an imprint of Taylor & Francis Group, an informa business

No claim to original U.S. Government works

ISBN-13: 978-1-03-209524-0 (pbk)
ISBN-13: 978-1-138-08292-2 (hbk)

Typeset by Apex CoVantage, LLC

Library of Congress Cataloging-in-Publication Data
Names: Szwedzicki, Tadeusz, author.
Title: Rock mass response to mining activities : inferring large-scale rock mass failure /
    Tadeusz Szwedzicki, Independent Consultant in Mining Geomechanics, Sorrento,
    Australia.
Description: Leiden, The Netherlands : CRC Press/Balkema, [2018] |
    Series: Geomechanics research | Includes bibliographical references and index.
Identifiers: LCCN 2018015611 (print) | LCCN 2018016469 (ebook) |
    ISBN 9781315112336 (ebook) | ISBN 9781138082922 (hardcover : alk. paper)
Subjects: LCSH: Ground control (Mining) | Mine accidents—Risk assessment. |
    Rock bursts. | Rock mechanics.
Classification: LCC TN288 (ebook) | LCC TN288 .S985 2018 (print) | DDC 622/.28—dc23
LC record available at https://lccn.loc.gov/2018015611

Contents

Preface		ix
About the author		xi

1 Introduction | | 1

2 Factors affecting rock mass response to mining | | 5
2.1	Geotechnical factors affecting rock mass response	5
2.2	Mining factors affecting rock mass response	6
	2.2.1 Rock mass response to development blasting	9
2.3	Effect of mining scale on response of rock mass	10
	2.3.1 Mining geometry	11
	2.3.2 Rock mass response at excavation scale	13
	2.3.3 Rock mass response at level scale	14
	2.3.4 Rock mass response at mine scale	17

3 Case studies of rock mass response to underground mining | | 21
3.1	Case studies of surface crown pillar collapse	21
	3.1.1 Coronation mine	23
	3.1.2 Chaffers Shaft area	24
	3.1.3 Iron King mine	25
	3.1.4 Nobles Nob mine	25
	3.1.5 Perseverance Shaft area	29
	3.1.6 Prince of Wales mine	30
	3.1.7 Scotia mine	32
	3.1.8 Warrego mine	32
	3.1.9 Analysis of surface crown pillar collapses	34
3.2	Case studies of rockbursts and outbursts	36
	3.2.1 Case study of precursors to rockburst	36
	3.2.2 Case studies of precursors to gas and rock outbursts	37
	3.2.3 Case study of gas outburst	38
	3.2.4 Case study of geothermal outbursts	39

3.3	Case studies of uncontrolled caving and pillar collapses	40
	3.3.1 Discontinuous subsidence over a caving mine	41
	3.3.2 Propagation of a caving zone	42
	3.3.3 Collapse of the rock mass over caving area	45
	3.3.4 Collapse of a roof due to pillar failure	45
	3.3.5 Collapse of pillars in a colliery	46
	3.3.6 Progressive collapse in room-and-pillar trona mine	47
3.4	Case studies of damage to underground mining infrastructure	48
	3.4.1 Enlargement of orepasses	48
	3.4.2 Ground movement in a decline	57
	3.4.3 Shaft collapse	59

4 Case studies of rock mass response to surface mining   **61**

4.1	Open pit slope failure due to underground mining	61
4.2	Slope failure along geotechnical structures	62
4.3	Collapse of a highwall in an open pit	64

5 Case studies of inundations   **65**

5.1	Water inrush into a colliery	66
5.2	Tailings inrush into an underground mine	66
5.3	Backfill liquefaction and inrush into a mine	68
5.4	Mud inrush resulting from collapse of a crown pillar	68
5.5	Instability of waste rock tip	69
5.6	Instability of tailings dam	70
5.7	Progressive failure of coal refuse dam	71
5.8	Failure of tailings dams triggered by earthquakes	72

6 Effect of discontinuities on the initiation of failure process   **75**

6.1	Modes of failure of rock samples	75
	6.1.1 Simple extension	75
	6.1.2 Multiple extension	76
	6.1.3 Multiple fracturing	76
	6.1.4 Multiple shear	76
	6.1.5 Simple shear	76
6.2	Effect of discontinuities on strength of rock samples	79

7 Behaviour of rock mass prior to failure   **81**

7.1	Pre-failure warning signs	81
	7.1.1 Indicators	81
	7.1.2 Precursors	83
	7.1.3 Triggers	92
7.2	Sequence of precursors to rock mass failure	93

7.3   Geotechnical risk management   94
7.4   Monitoring of precursory behaviour   95

8 Rock mass behaviour during failure   **99**
8.1   Case studies on onset of failure   99
8.2   Case studies on duration of failure   100
8.3   Progressive damage to excavations under high mining-
      induced stress   101
      8.3.1   Case study of stress transfer through remnant pillar   104
      8.3.2   Case study of stress transfer through the down-dip
              abutment   105
      8.3.3   Case study of stress transfer through compacted
              caved rocks   106

9 Post-failure rock mass behaviour   **111**
9.1   Case studies of post-failure behaviour   111
9.2   The effect of mining geometry on post-failure behaviour   115

10 Modes of failure of rock and rock mass   **119**
10.1   Modes of failure at a sample scale   120
10.2   Mode of failure at a local scale   123
10.3   Mode of failure at a mine scale   130

11 Behaviour of fragmented ore   **133**
11.1   Ore flow – a case study from a block caving mine   134
       11.1.1   Rock material in a draw zone   136
       11.1.2   Diameter of a draw zone   137
       11.1.3   Change in fragmentation due to draw   138
       11.1.4   Effect of ore draw rate on damage of drawpoints   139
11.2   Behaviour of fragmented ore – a case study from a sublevel
       caving mine   140
       11.2.1   Ore fragmentation and size distribution   140
       11.2.2   Ore recovery   141
       11.2.3   Primary recovery   141
       11.2.4   Secondary recovery   142
       11.2.5   Tertiary and quaternary recoveries   143
       11.2.6   Effect of rock mass properties on fragmentation   143
       11.2.7   Reduction of fragment size with tonnage drawn   143
       11.2.8   Draw management practices   146
11.3   Analysis of a diameter of a draw zone from two case studies
       of caving mines   148
11.4   Behaviour of fragmented ore in orepasses   149
       11.4.1   Types of hang-ups   149

	11.4.2 Formation of hang-ups	150
	11.4.3 Hang-up prevention	152

12 Mitigation of rock mass response through geotechnical quality assurance   155

12.1	Quality in ground control	156
12.2	Quality assessment	157
12.3	Quality assurance in ground control management system	158
	12.3.1 Responsibilities and authority	158
	12.3.2 Compliance with legislation	159
	12.3.3 Competency and training	159
	12.3.4 Communication and reporting	159
	12.3.5 Document control	159
12.4	Quality assurance in geotechnical planning and design	159
	12.4.1 Data collection and analysis	160
	12.4.2 Geotechnical planning	160
	12.4.3 Ground Control Management Plan	160
	12.4.4 Mine closure	161
	12.4.5 Geotechnical design	161
	12.4.6 Approval system	161
	12.4.7 Feedback and follow-up	161
12.5	Quality assurance in ground control activities	162
	12.5.1 Drilling	162
	12.5.2 Blasting	162
	12.5.3 Maintenance of excavations	164
	12.5.4 Performance of ground support	166
12.6	Quality assurance in geotechnical inspection and monitoring	168
	12.6.1 Inspection	168
	12.6.2 Instrumentation	169
	12.6.3 Monitoring of rock mass performance	169

References	171
Index	177
Book series page	180

Preface

The concepts covered in this book offer significant benefit to professionals with expertise in geotechnical engineering, mining engineering and/or geology aiming to understand rock mass behaviour associated with mining activities.

To predict a geotechnical event in a mine, it is critical to specify type of damage, location of damage, severity and time. This book focusses on two elements: (1) inferring type of damage and (2) specifying location of potential damage.

The case studies presented in the book demonstrate that in all cases the severities of geotechnical mining disasters were unimaginable (numerous fatalities and loss in billions of dollars), and that timing of the events depended on internal or external triggers which were seldom predictable.

The book discusses geotechnical indicators and warning signs of impending or progressive damage, collapse or rock mass failure which together with analysis of mining parameters could provide guidelines for prevention or mitigation of damage to mining excavations.

After finishing the book, you should be able to read rock mass behaviour and should be able to detect the tell-tale signs of impending rock mass damage.

T. Szwedzicki

About the author

Dr Tadeusz Szwedzicki is an internationally recognised expert in geomechanics of underground mining methods. He has over 40 years of mining experience working in mining production, research and development, and consulting. His experience has been gained working for some of the world's largest mining companies like PT Freeport (Indonesia), ZCCM (Zambia), Anglo American Corp (Republic of South Africa), and WMC and BHP Billiton (Australia). He held academic positions at Western Australian School of Mines, Curtin University of Technology and University of Zimbabwe. His experience includes government position at the Northern Territory Department of Mines and Energy where he was appointed the Government Mining Engineer. He also worked for the Government of Papua New Guinea as Mineral Resources Advisor. He is a recipient of the Silver Medal awarded by the Institution of Mining and Metallurgy, London, and a recipient of a Fulbright scholarship, USA. He has authored over 70 papers including in the *International Journal of Rock Mechanics*, *Transactions of the Institute of Mining and Metallurgy*, and proceedings of international conferences. He is an independent consultant specializing in geomechanics of mining methods.

Chapter 1

Introduction

All failures resulting from human activities are predictable.

anonymous

Just a few days after I started working at a deep underground hard rock mine as a geotechnical engineer, a double fatality happened. Two miners carrying a heavy pump up a decline were killed by a rockfall. Investigation that took place described the accident as "unpredictable" and stated that the support was adequate for the prevailing conditions. One day, soon after the rockfall, when I was walking up the decline, I noticed that at that particular spot there were some wet patches at the back and some water trickling at one sidewall. After a few walks, I noticed the width of the decline was about half a metre wider than in other places, which wasn't clearly visible because of irregular walls caused by blasting. After thorough inspection, I realised that the rockfall was in a fault zone. The rock mass within the fault (which was about 5 m wide) was similar to the neighbouring country rock but was highly jointed. All these observations made me think if that rockfall was really unpredictable. That accident was a reminder that the paramount objective of mining geomechanics is to ensure safety of mining personnel. This marked the beginning of my geotechnical quest – to determine whether we are in fact capable of reading the signs of rock mass response to mining . . . and therefore, predicting potential geotechnical hazards such as failure of rock around mining excavation or even potential disasters like ground collapse, rockburst or inundation?

My quest has been pursued globally in mines in Europe (Poland), Africa (Zambia, Zimbabwe, Republic of South Africa), and Australasia (Papua New Guinea, Indonesia, and Australia).

***

Identification of mining geotechnical hazard shall be based on observations and monitoring of rock mass behaviour under mining-induced stress and on analysis of mining parameters. This book deals with the challenge and covers two intertwining topics: factors governing rock mass response to mining activities and rock mass behaviour before, during, and after failure of rock and rock mass. These two topics are supported by numerous case studies and by discussion on modes of rock and rock mass failure.

Various manifestations of rock mass behaviour as response to mining can be identified in all phases of mining activities i.e. during development work and during production activities in stopes (Chapter 2).

A number of case studies are reviewed on large-scale failures and disasters in underground mines and open pit mines, on instability of tailings dam and waste material dumps and on inundation of mining areas. The reviews of the case studies are focused on rock mass response during the progressive damage, failure and post-failure rock mass response. The case studies on underground mining, in Chapter 3, include the following:

- surface crown pillars collapse above underground mining excavations,
- rockbursts, gas outbursts and geothermal outburst,
- uncontrolled caving and pillar collapse, and
- damage to underground infrastructures (large excavations, accesses and shafts).

Chapter 4 provides case studies on slope failure in open pits, including the following:

- failure due to underground mining below,
- failure along geotechnical structures, and
- collapse of a highwall.

Inundation and liquefaction are presented in Chapter 5 for the following:

- water inrush into a colliery,
- tailings inrush into an underground mine,
- backfill liquefaction and inrush,
- mud inrush following pillar collapse, and
- instability of waste rock dumps and tailings dams.

Damage can be stress induced or structurally controlled (or in a prevailing number of cases a combination of these). However, the fracture propagation that is caused by stress increase is instigated on microfractures. The detection and their effect on rock sample mode of failure is discussed in Chapter 6.

The ability to recognize pre-failure rock mass behaviour may result in predicting and averting the potential for geotechnical damage. Precursors to mining failures (like indicators, warning signs and triggers) are reviewed in Chapter 7.

Chapter 8 covers rock mass response at the onset of failure and duration of failure process. The chapter also provides case studies of progressive damage.

After failure the rock mass exhibits residual post-failure behaviour. The behaviour can be re-occurring and can last a long time. This must be considered when re-entering affected areas i.e. for post-event recovery or continuation of mining, as discussed in Chapter 9.

Modes of rock sample failure and modes of rock mass failure on local and mining scale are reviewed in Chapter 10.

Behaviour of fragmented ore and waste rock can affect rock mass response to mining. Rock mass fragmented after blasting may provide confinement to the surrounding rock mass. Compacted broken rock can transfer stresses that may result in ground deterioration around drawpoints and crosscuts, as discussed in Chapter 11.

Case studies demonstrate repeatedly that observation of rock mass response and timely implementation of ground control practices can mitigate the effect of stress changes leading to damage. However, the critical mitigation factor is the implementation of geotechnical quality assurance, as discussed in Chapter 12.

***

Mining activities result in change in mining-induced stress. Changes in stress around mining excavations result in changes in the behaviour of the rock mass, which in turn may lead to mining disasters due to damage, failure and consequent collapse of the rock mass. Mining disasters may result in multiple fatalities, environmental damage and severe financial losses. The type and scale of response depends on *in situ* and mining-induced stress, structural features and rock mass strength, as well as mining geometry and the scale of mining operations.

Once the first signs of stress are observed, such as, cracking or fracturing of rock mass or damage to ground support, the excavations start to deteriorate. Damage to the rock mass can pose various geotechnical hazards like collapse of ground (fall of ground, crown or protective pillar collapse), seismic activities, slope instability, and inundation or instability of backfill, mine tailings or waste rock. The deterioration can progress linearly or exponentially i.e. deterioration begins slowly but then accelerates towards eventual closure or collapse. Rapid and violent failures of large-scale geotechnical mining structures cause significant safety hazards, material damage and interruption to or even cessation of mining activities. It is vital to acknowledge that all mining companies are vulnerable to such geotechnical events.

In small mining operations in low-stress environments, the rock mass response is hardly visible and such excavations have a very long life. However, with larger mining operations, especially in deep mines, the response can indicate mining-induced geotechnical hazards. Each rock mass failure is preceded by a precursory manifestation of rock mass behaviour.

Analysis of case studies shows that the rock mass responses can escalate in scale and finally end up in progressive damage, failure and/or collapse. Response to change of stress around mining excavations can be noticed long before failure. During the failure, different modes of rock mass failure take place and, finally, there is post-failure (residual) behaviour. Assessment of post-failure behaviour is required when making a decision on the timing of entering rescue teams, continuation of mining operations near affected areas, and even the surface utilization of a collapsed mine.

Ability to recognise indicators and warning signs may result in predicting or averting the potential for geotechnical failure and thus avoiding substantial losses. Unfortunately, precursors are not always recognised before the occurrence but are rather recalled in hindsight, during investigations into the disasters. In many occurrences, geotechnical failures were classified as "accidental", "occurrence without precedent", "sudden failure without warning", "never anticipated or foreseen" or "unexpected" – yet on scrutiny were found to be not completely unpredictable. Instead, they could have been averted or at least the effects of failure could have been mitigated. Such failures often exceed engineering expectations of rock mass behaviour due to the large scale and severity of damage, which may be one of the reasons why they were often considered unexpected.

A variety of deficiencies may arise during the planning and design stages, and the most common are caused by incorrect siting of the development and by designing excavations of inappropriate size and shape. Damage to rock mass structures like pillars, stopes, chambers, magazine and secondary developments could be progressive or violent and may end up in closure or collapse. Mining history has clearly demonstrated disasters involving the collapse of pillars due to high extraction ratio, water or tailings inrush into mines as a result of the incorrect siting of surface water reservoirs and tailing dumps. Shafts have been abandoned because of damage to linings due to deformation caused by the unsatisfactory design of the shaft pillars.

When rock mass failure is accompanied by substantial uncontrolled rock movement, it is referred to as a collapse, for example, discontinuous subsidence, caving, rockfalls, slope instability or pillar disintegration. If rock mass failure is accompanied by an abrupt large energy release, it is referred to as a rockburst. If accompanied by abrupt large gas release, it is referred to as a gas outburst. When accompanied by a large increase in water inflow, it is referred to as inundation.

In large tailings dams or waste rock dumps, the failure takes place by slippage or flow of liquefied material. This is referred to as waste material instability.

Even with all the indicators and warning signs, mining companies seldom see the geotechnical failures coming. Even the most sophisticated and well-managed operations are frequently caught unaware by disastrous events – events that could have been anticipated and prepared for.

Anticipating and avoiding geotechnical events requires understanding of rock mass response to mining activities. Failure to do so will leave a company vulnerable to potentially devastating events.

Chapter 2

Factors affecting rock mass response to mining

Geotechnical failures are events that should be seen coming.

anonymous

Understanding factors affecting ground behaviour and prediction of rock mass response allows for assessment of vulnerability of mining infrastructure and mineral extraction processes. Once rock mass responses to mining are identified and risks determined, it is possible to implement appropriate mitigation actions. Understanding and foreseeing rock mass response and behaviour is needed for the mine design process, to select mining methods and to apply ground control techniques to ensure safe and efficient mining practices.

The consideration deals primarily with rock mass response where damage around mining structures occurs and where confining stresses are very low or tensile.

Under stress (which can be mining-induced or externally triggered by blasting or seismic events) a rock mass is subject to damage. Although the damaged rock mass may transfer stress and maintains its integrity, there is always risk of failure and collapse. Geotechnical failure of mining structures is defined as fracturing or disintegration resulting in loss of bearing capacity and loss of ability to perform its function. Although rock mass failure process is a function of rock properties, structural features and mining geometry, the failure itself is driven by stress changes. The loss of bearing capacity happens because of uncontrolled ground movement or energy release. When failure is accompanied by substantial discontinuous displacement of rock, it is referred to as rock mass collapse. Another form of rock mass response is progressive deterioration. Progressive deterioration takes place when the rock mass behaves in a ductile way. During such deterioration excavations change their shape without failure i.e. are able to continuously transfer stress until such deformation is achieved that a new stress balance is achieved. Mode of failure is defined as a manner, form or mechanism of rock or rock mass fracturing leading to failure under induced stress.

2.1 GEOTECHNICAL FACTORS AFFECTING ROCK MASS RESPONSE

Factors contributing to rock mass response leading to damage are structural features and rock mass mechanical properties, and *in situ* and mining-induced stress (Fig. 2.1). The figure also incorporates the role of failure criteria and refers to ground control techniques. Rock mass response can indicate stability or instability. Stability refers to open span of

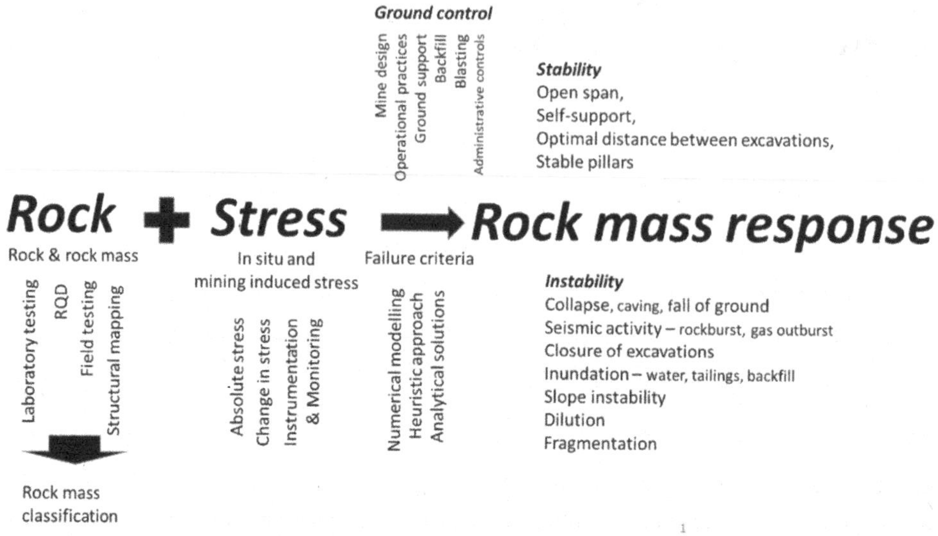

Figure 2.1 Geotechnical input to determine rock mass response.

excavations, self-support, optimal distance between excavations, stable pillars, etc. Instability, for example, can result in ground collapse, rock mass fragmentation, seismic activity, closure of excavations, slope movement, inundation, and failure of tailings storage facilities.

Two general modes of damage can be distinguished: structurally controlled gravity-driven and stress-induced failure with spalling or slabbing (or any combination of them). Structurally controlled modes of failures are most frequently observed at shallow depths, and stress-induced failure is commonly found at greater depth. At shallow depth, slip along discontinuities or shearing of the rock matrix dominates the failure process, while at depth stress-induced fracturing is most common (Kaiser, *et al.*, 2000).

Mechanical properties are determined by laboratory testing of rock samples. Rock mass is described by the Rock Quality Designation/Fracture frequency number, field testing and structural mapping. All these factors allow for geotechnical classification of the rock mass (Brady & Brown, 1993). Blasting, stress fracturing and water often reduce rock mass properties in the vicinity of mining excavations. Rock mass behaviour, in each geotechnical domain, also depends on combination of often contiguous very poor and good ground domains. Large structural features like faults, folds and joints can control the mode of failure. However, damage is often instigated by stress concentration around microfractures (see Chapter 6).

2.2   MINING FACTORS AFFECTING ROCK MASS RESPONSE

Stress is considered as superimposed *in situ* stress and mining-induced stress and is determined by measurement of the absolute values of stress and by measurements of stress changes. Although *in situ* stress doesn't change during the life of a mine, the mining-induced stress changes during mining activities. Mining-induced stress depends on mining geometry,

geometry of mining excavation and compaction of fragmented rocks. Mining geometry, when referring to mine production, includes the extraction ratio, pillar width-to-height ratio, sequencing of extraction and ground control measures such as backfill or support. Geometry of mining excavations includes shape and open span, pillar size, interaction between neighbouring excavations, etc. When using caving, open-stoping or shrinkage methods, there is one more factor – broken rocks. Rock mass fragmented after blasting, when left in stopes, cave zones or orepasses, provides confinement to the surrounding rock mass that may be affected by abutment stress. Compacted broken rock can transfer stresses that may result in ground deterioration around drawpoints and crosscuts.

Failure criteria like numerical modelling, heuristic methods or analytical solutions determine if the rock mass is susceptible to failure i.e. its response results in stability or instability but is not used to determine the mode of failure.

Predicted instability is indicative of rock mass collapse, fall of ground, seismic activities, dilution, fragmentation and rock mass damage; when instability is foreseen, ground control measures such as appropriate mine design, ground support or backfill must be considered.

Various manifestations of rock mass behaviour as response to mining can be identified in all phases of mining activities. Damage to the rock mass can be described by defining the extent (local or mine scale), location of the damage (pillar, floor or back or larger scale), mode of failure (tension, shear or coupled) and rock mass response (brittle or ductile).

Rock mass response can be controlled during design and planning stage, development stage and production (Fig. 2.2; Szwedzicki, *et al.*, 2007).

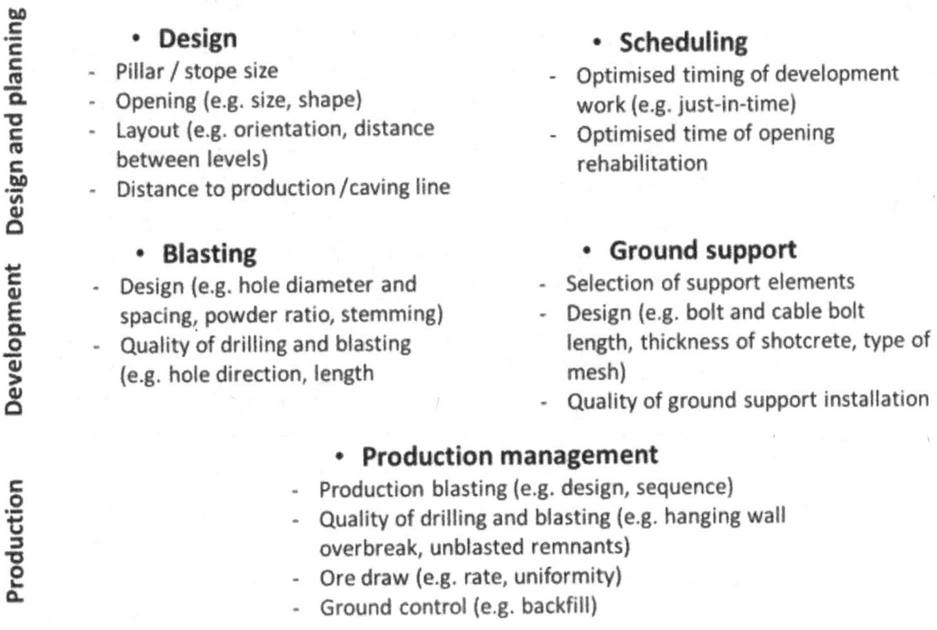

Design and planning

- **Design**
  - Pillar / stope size
  - Opening (e.g. size, shape)
  - Layout (e.g. orientation, distance between levels)
  - Distance to production / caving line

- **Scheduling**
  - Optimised timing of development work (e.g. just-in-time)
  - Optimised time of opening rehabilitation

Development

- **Blasting**
  - Design (e.g. hole diameter and spacing, powder ratio, stemming)
  - Quality of drilling and blasting (e.g. hole direction, length

- **Ground support**
  - Selection of support elements
  - Design (e.g. bolt and cable bolt length, thickness of shotcrete, type of mesh)
  - Quality of ground support installation

Production

- **Production management**
  - Production blasting (e.g. design, sequence)
  - Quality of drilling and blasting (e.g. hanging wall overbreak, unblasted remnants)
  - Ore draw (e.g. rate, uniformity)
  - Ground control (e.g. backfill)

Figure 2.2 Mining factors affecting rock mass response (Szwedzicki *et al.*, 2007).

For every stage of mining activities, different operational controls can be used to mitigate the resulting vulnerabilities. During mine planning and design the following mining factors should be considered: pillars and stope dimensions, excavation size and shape, mining lay-out and sequence (e.g. orientation in relation to *in situ* stress and major structural features, distance between levels), distance of secondary excavations to a production/caving front, optimised timing of development, ground control measures, the maximum open span of all excavations, etc.

For development, scheduling support work and blasting play important roles in control-ling rock mass response. For ground support, the suitability of all support and reinforcement elements should be considered: various bolts and cable bolts, wire mesh of different type and shotcrete. It should be noted that quality assurance and quality control of blasting and support work plays an important role in mitigating damage to the rock mass. For efficient blasting, the following factors should be considered: hole diameter, burden, powder factor and spacing between holes. As scheduling is concerned, optimised timing of development work (just in time) and optimised time of rehabilitation of excavations is important.

During production in stopes, ground behaviour can be managed by quality of production drilling and blasting, by sequencing of blasting and ore draw and by using backfill and other ground control measures.

Rock mass response to mining activities depends on the mining methods. For open stop-ing and room and pillar, the most common responses of the rock mass are failure of hanging-wall, pillar damage, damage to drawpoints and access developments (due to stress caused by remnant and stress transfer) and also ore dilution. An example of the hangingwall collapse despite heavy ground support is given in Fig. 2.3.

Figure 2.3 Failure of a reinforced hangingwall in an open stope.

For block and sublevel caving, the most common challenges are damage to the access opening due to remnants caused by poor-quality blasting, damage due to ore compaction, damage to peripheral excavations due to stress transfer in front of the caving face (abutment stress) and increases in cave loading due to poor draw practices. During the production stage, high mining-induced stress causes ground movement resulting in crosscut convergence or instability, especially in weak rocks.

During the total extraction the rock mass response is manifested by ground subsidence. The subsidence that affects surface and subsurface infrastructure could be controlled and mitigated by selective mining and backfill.

2.2.1 Rock mass response to development blasting

Development of mine openings not only changes stress distribution but also results in rock mass damage due to blasting. The damage in the form of fractures results in reduction of mechanical properties of the rock mass and formation of loose rocks around mining excavations. Fractured and loose rocks are controlled by scaling (barring down) and ground support. After development blasting all loose rocks must be scaled down before the ground is reinforced by shotcrete, bolts and mesh. Loose rocks that are not scaled precisely cause financial losses during shotcreting. Small rock fragments, detaching under the weight of freshly sprayed shotcrete, can fall off resulting in increased shotcrete consumption and increased turnaround time.

A degree of blast damage was assessed using hydroscaling (water jet scaling) to remove loose rocks from the walls of new developments after blasting. The waterjet scaling process is shown in Figure 2.4. Time for waterjet scaling of ground exposed after 3 m cut varied from

Figure 2.4 Hydroscaling of a development face (Szwedzicki *et al.*, 2007).

Figure 2.5 Loose rocks lying on a tarpaulin spread on the floor after hydroscaling (Szwedzicki *et al.*, 2007).

2 to 16 minutes. The best results were obtained when water was ejected from a nozzle on the boom of a shotcreting machine under pressure from 10 MPa to 12 MPa with water flow rate of 250 l/min.

Results proved that when the smooth blasting technique was used (i.e. most of the half barrels were visible), the volume of scaled-down rocks was minimal (less than 0.5 tonne) per 3 m cut. Figure 2.5 shows loose rocks lying on a tarpaulin after hydroscaling of a development heading. In the case of blast holes drilled on larger than 0.6 m spacing, with deviating holes or in poor ground conditions, the volume of scaled-down rocks increased to over 3 tonnes per cut (Dunn *et al.*, 2006).

2.3   EFFECT OF MINING SCALE ON RESPONSE OF ROCK MASS

It is practical to consider two scales of rock mass response that might require ground control measures:

- "local" scale ground response indicating rock damage in the immediate vicinity of excavations, and
- "mine" scale ground response (also known as large, global or regional scales) indicating rock mass damage.

The local scale of response includes "excavation" and "level" scales of ground control.

Excavation scale behaviour affects rocks around mine openings (up to one radius of the opening from the excavation boundary) rather than the rock mass. Rock mass damage between one and three radii from the excavation boundary can be considered as being of level scale, for example, damage to small size underground infrastructure like accesses, drawpoints or workshops. Damage that started at excavation scale can change quickly to the level scale due to mining-induced stresses. In such a situation it is difficult to assess the scale, and for practical reasons both excavation and level damage is considered as local scale damage.

At an excavation scale, in response to development blasting, rock mass structure is damaged, and fractures can propagate, leading to formation of loose rocks on the periphery of openings. The rock failure is usually manifested as cracking, joint opening, spalling, slabbing, pillar rupture, rockfalls or strain bursts. The local response and its changes are relatively easy to observe and can be remedied by support and reinforcement.

Rock mass response to production activities can affect mining level accesses, mining infrastructure (e.g. shafts, orepasses, chambers, declines, magazines) and the surface.

At a level scale, rock mass response during mining activities is observed as instability of pillars, progressive failure of walls or closure of excavations. The most typical rock failure for this scale is pillar punching the hangingwall or footwall (which may result in floor heaving or roof guttering), breakouts around excavations, rockfalls, crack propagation (often between levels) or strain bursts. Under high mining-induced stress, failure can propagate for a large distance and finally result in mine scale failure. This phenomenon can take months or years. Changes in the level scale zone are more difficult to recognise but can be remedied and controlled by heavy support like cable bolts, thick shotcrete or steel arches. The controls to that response should be implemented during planning and design, by using continuously improved ground support and by rehabilitation.

Mine scale damage affects the rock mass through large parts of the mine and can extend many hundreds of metres, often to the surface. Mine scale level responses, which take place deep in the rock mass, are usually not easily recognisable and, once they take place, the risk of uncontrolled rock mass instability is high. At a mine scale, the response to total extraction is manifested as subsidence affecting the stability of mining infrastructure on the surface and underground. In certain mining situations, damage of rock mass structure may be desired, for example, subsidence above longwall panels or yielding pillars. The behaviour and instability are said to affect the rock mass through large parts of a mine. They are manifested in movement along shear zones, in closure of excavation, or as pillar and hangingwall collapse, surface subsidence and sinkhole formation, caving or large rockbursts.

The controls that could be implemented to ameliorate the damage are backfilling and changing extraction sequencing, developing bypasses and using heavy support.

It was observed that the predominant controlling factors at an excavation scale were structures and rock strength. The predominant factors at the level and mine scales are the extraction ratio and pillar width-to-height ratio.

2.3.1   Mining geometry

Mining geometry for underground mine design is often decided based on observations and experience and/or using guidelines derived from case studies. Rock mass stability depends on mining geometry – the size and shape of pillars left to support the hangingwall and the volume of rock material excavated. Mining geometry is commonly defined by two parameters: extraction ratio and pillar shape. The extraction ratio, defined as mined out area/(mined out area + pillar area), describes the open unsupported span between pillars. This parameter uses dimensions in two directions within the plane of the deposit, whereas the parameter of pillar shape describes the width:height ratio of a pillar. This second parameter takes into account dimensions of a pillar in the plane of the deposit and perpendicular to it. Although a higher extraction ratio is often associated with lower pillar width:height ratio, both parameters are independent.

Mining geometry determines mining-induced stress concentration and stress distribution in rock mass which in turn can result in changes in ground conditions. Under high stress, rock and rock mass may be damaged by fracturing, cracking, yielding, spalling, etc. Such damage may result in increased geotechnical hazards but not necessarily impede mining operations.

Mine design is based on assumptions of pre-failure behaviour of rock mass. Many methods of stability analysis, in their formulae or rock mass classifications, use the uniaxial compressive strength of rock samples as the stress limit, for example, for pillar design or stability assessment. From a geotechnical point of view, it is not only important that the pre-failure behaviour is known but also that the rock mass post-failure behaviour is considered.

Although it is well known that pillars transmit stress even though they fracture and are damaged, calculation of the pillar geometry is based on the strength and rock behaviour obtained on intact rock samples tested for pre-failure parameters. Additionally, it is usually assumed that damage is confined to a pillar and does not take into account damage to the floor or the back (hangingwall and footwall).

To determine the effects of mining geometry, underground investigations and observations and laboratory testing were conducted. Work on the effect of pillar width:height ratio on post-failure behaviour has been conducted on coal samples (Wagner, 1984;; Madden, 1989), and hard rock pillar samples (Starfield and Wawersik, 1968). To determine the post-failure behaviour of samples which represented pillars and the surrounding rock mass, testing was carried out on shaped drill cores (Mendes and Da Gama, 1972).

To determine the relationship between mining geometry and behaviour of pillars, underground studies were conducted and complemented by pillar modelling.

Rock behaviour under compressive stress was modelled on:

- samples of various width:height ratios (discs of different height) (Ormonde and Szwedzicki, 1993).
- large diameter core shaped to represent pillars of different geometry within the surrounding hangingwall and footwall (Kilpatrick and Szwedzicki, 1994). To represent pillar and surrounding rock mass stability, dumb-bell-shaped samples consisted of a pillar in the centre and two ends which represented the hangingwall and footwall respectively. The reduction in diameter at the central part of the sample was representing mining activity and expressed as an extraction ratio.
- core samples with openings representing mining excavations. By changing the number of horizontal openings, the geometry was varied, resulting in a changing pillar width:height ratio and extraction ratio.

Pillars were classified according to their width:height ratio. Pillars with a width:height ratio of less than 2 were classified as slender, with a ratio between 2 and 4 as regular, and pillars with the ratio larger than 4 as squat. An extraction ratio less than 40% was classified as low, between 40% and 80% as typical, and larger than 80% as high.

2.3.2   Rock mass response at excavation scale

Instability at an excavation scale, which takes place predominantly in the midheight part of pillars, was recorded in samples where the pillar width:height ratio was less than 2 and the extraction ratio more than 80% i.e. slender pillars and a high extraction ratio. Such mining geometry is encountered when a deposit is extracted by room-and-pillar mining, leaving small shaft pillars and small pillars between crosscuts. In all these cases, damage initiated at the midheight part of pillars was manifested by fracturing, spalling, slabbing or shearing. This type of damage eventually propagated to encompass the entire pillar.

Reduction of pillar load-bearing capacity was predominantly associated with progressive fragmentation and formation of fractures extending through the pillars. Figure 2.6A shows progressive deterioration of a reinforced pillar. Similar behaviour was noticed on slender pillar models which exhibited brittle behaviour (Fig. 2.6B).

Figure 2.6A Progressive fragmentation of a pillar in high extraction area.

Figure 2.6B Spalling of a model pillar (width:height ratio 1.5) (Szwedzicki, 2000).

This mode of failure is often seen in pillars that are coming under high mining-induced stress, when production activities approach the pillars. Although slender pillars usually fail progressively, they can fail violently, in a rupture mode, either at or immediately following the attainment of peak strength.

2.3.3   Rock mass response at level scale

With an increase in width:height ratio to greater than 2 but less than 4 and a decrease in extraction ratio from below 80% but still above 40%, the rock mass responds differently and instability changes from predominantly an excavation a scale to the level scale. Such mining geometry is encountered when a deposit is partially extracted.

Under mining-induced stress, pillars deteriorate and the damage extends to the hangingwall. As fractured and spalled off material accumulated at the periphery of a pillar, the confining stress built up and dilation of fractured pillars is restricted. For such conditions, the slope of the post-failure curves trended in a negative direction, i.e. the load-bearing capacity was reduced as the pillar strain increased. In mining practice, pillars of that ratio are designed as yielding pillars.

Pillars with a ratio above 2 can induce failure to the hangingwall or footwall. Local stability issues include failure of the immediate back (hangingwall) or floor (footwall) heaving between pillars. The failure through the hangingwall or footwall was due to a punching effect (Fig. 2.7A). Punching of the hangingwall may result in so-called guttering with rock fragments falling into mining excavations and punching in the footwall may result in floor heave. Both effects are highly detrimental to mining activity. Figure 2.7B shows a model sample that failed in axial fracturing and/or shear mode, caused by a pillar punching into the hangingwall.

Figure 2.7A Coal pillar punching into the back.

Figure 2.7B Failure through the pillar and hangingwall part of a model pillar (width: height ratio 2).

Figure 2.8A Damage around an excavation through floor heave, shoulder spalling and formation of an extension crack in the centre of the excavation.

Figure 2.8B Failure through the pillar and the hangingwall and footwall parts of a model of a regular pillar (Szwedzicki, 1989a).

Damage to regular pillars can also include fragmentation of the exposed walls of hangingwall lowering and footwall heave. Figure 2.8A shows fragmentation of exposed walls of pillars forming so-called dog earing and footwall heave.

Such rock mass damage can be accompanied by a vertical extension cracking in the centre of the floor. That type of damage is visible on model pillars where openings were defining pillars size and shape (Fig. 2.8B).

2.3.4   Rock mass response at mine scale

Rock mass deterioration at a mining scale usually starts on a local scale and progresses through the country rock. Such mode takes place for pillars with a width:height ratio more than 4 (squat pillars). The damage takes place in the pillars and through the surrounding rock mass.

For such a ratio, pillars start failing on the periphery in a distinct circular fracturing pattern and the failure progresses to the core of the pillar. The extent of fracturing depends on the stress level. In many observed underground pillars, the peripheral damage progresses and stabilises, preventing total pillar collapse. The depth of the fractured zone was investigated by drilling boreholes into pillars. Figure 2.9A shows open cracks close to the collars of the borehole. The recovered core showed that the depth of the fracturing, depending on stress distribution, can vary on different sides of the pillar. For the specific site, the recorded depth varied

Figure 2.9A Visible open cracks in the wall of a squat pillar.

Figure 2.9B Concentric fractures with a solid core in the centre (width: height ratio greater than 4).

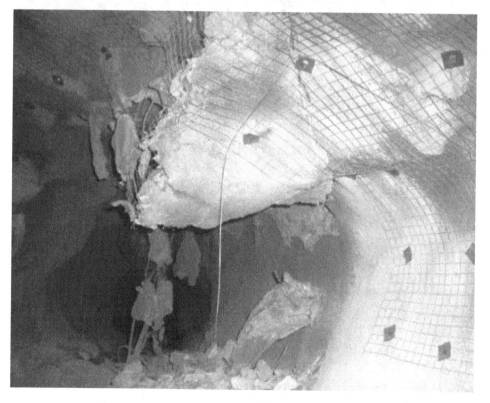

Figure 2.10A Damage to the hangingwall at the shoulder of a squat pillar.

Figure 2.10B Mine scale failure through a hangingwall part of a model.

from 0.8 m to 1.2 m. The core part of the sample remained solid, unfractured rock. Similar behaviour was observed on model pillars with width:height ratio larger than 4 (Fig. 2.9B).

An example of rock mass damage above squat pillars is given in Figure 2.10A and B. In Figure 2.10A, the damage took place at the shoulder of a pillar and the potential fall of ground was restricted by ground support. When testing a shaped squat model pillar, a large shear failure, which ran through the hangingwall of the sample, led to a rapid loss of strength (Fig. 2.10B).

Stress-strain relationship for samples with a width:height ratio of 4.0 (squat pillars) confirmed that geometrical configuration of a pillar had an effect on post-failure behaviour. With a pillar width:height ratio greater than 4.0, the slope of post-failure curves after reaching peak strength (defined as stress at the limit of elastic sample behaviour) became positive and the load-bearing capacity began to increase. Squat pillars disintegrated progressively which allowed for substantial post-failure deformation. With progressive damage, there was a reduction in the effective cross-sectional area of the specimen and an increase in confining forces. An explanation for this behaviour is that as width:height ratio increases, fragments do not fall away from the specimen but remain in place and the lateral segments restrain each other, producing radial confining forces.

However, in mining practice, these fragments are often removed so that confinement is not large enough to maintain the constant stress during deformation.

Underground observations of pillar response to mining and laboratory testing on modelled pillar show that modes of pillar failure vary for different geometry of pillars. Slender pillars fail and disintegrate while the hangingwall and the footwall remain relatively undamaged. Such a failure results in an excavation scale damage. For regular pillars the damage can extend to the hangingwall and can be classified as level scale. For squat pillars the fractures propagate far into the hangingwall and the footwall resulting in mine scale damage.

Chapter 3

Case studies of rock mass response to underground mining

Mining structures can suffer because of rock mass damage. These structures, when under stress (which can be mining-induced or brought about by external conditions) as result of damage, are subject to changes in mechanical properties. Structures can suffer damage even though they may maintain partial integrity and perform their function. The damage process starts with a failure initiation phase, progresses through a propagation phase and ends with rock mass collapse, inundation, and rockbursts and gas outbursts.

> *Collapse of the rock mass.* Rock mass, because of mining activities, can suffer structural damage, fail and undergo substantial uncontrolled mass movement. This type of rock mass failure can take the form of discontinuous subsidence, caving, stope collapse or the disintegration of pillars. Rock mass collapse is expected to continue until the stress transfer around the collapse zone reaches equilibrium. Some rock mass collapses involved uncontrolled movement of rock mass in excess of several million tonnes.

> *Inundation.* Water or liquefied granular material, like tailings or waste rock, constitute a very serious hazard for underground and open pit mines. The pages of the history of mining activities are full of mine disasters caused by inrush of water, tailings, waste rock or backfill. Advances in mining technology and mining legislation have resulted in a substantial reduction of water inrushes, but such occurrences continue to happen. Water inrushes are classified as geotechnical failures because they are generally caused by the failure of protective pillars between the body of water and mining excavations.

> *Rockbursts and gas outbursts.* The intensity of rockbursts and gas outbursts increases as underground mining goes deeper. Despite substantial research into the mechanisms of these phenomena, it is generally accepted that the prediction of such events in terms of magnitude and time has not been successful. The most frequently reported precursors involve patterns of seismic and acoustic emission.

3.1 CASE STUDIES OF SURFACE CROWN PILLAR COLLAPSE

Surface crown pillars form a protection for the underground workings. Hard rock mining of mineral deposits often creates large underground voids which, in time, may lead to ground deterioration and collapse. Such collapses may progress to the surface causing continuous or discontinuous subsidence, Continuous subsidence takes place when deposits are mined over large areas, for example coal mining using longwall mining method. Discontinuous subsidence take place in form of sinkholes over large mining voids and usually takes place in hard rock mining. In many cases buildings, mine plants, roads and railways built over areas of past mining activity can be classified as areas of potential ground movement or collapse. Although caving of a crown pillar over hard rock mines is relatively rare, it creates hazardous conditions, which may jeopardise the lives of workers and could result in closure of a mine (Carter and Miller, 1995). Uncontrolled movement of large volumes of caved ground can be dangerous for people and equipment both on the surface and underground. Heavy losses of equipment have been reported due to substantial ground subsidence. Sinkholes can connect the surface with underground excavations and may facilitate water and tailings inrushes or flooding of the mine. The seriousness of such collapses is demonstrated by the fact that several mines ceased production because of the collapse. The main factors contributing to surface crown pillar collapses are as follows:

- the geometry of underground openings and of a crown pillar (open unsupported span and the thickness of the crown pillar),
- high extraction ratio (small pillars),
- inclined deposits (so that collapsed material can rill down the stopes preventing self-supporting of the hangingwall), and
- mechanical properties of the rock mass (e.g. oxidised rock material forming the crown pillar).

The following terms and definitions are used.

Sinkhole: a cylindrical or conical depression caused by discontinuous subsidence of rock mass over mined-out excavations.

Depth of a sinkhole: the vertical distance from the ground surface to the surface of the collapsed material.

Thickness of a surface crown pillar: the vertical distance between the highest point of an underground excavation and the ground surface. Unless otherwise specified, the "thickness" may include unconsolidated surface material and soil.

Width of an underground pillar: the pillar dimension in the strike direction, parallel to the hangingwall or footwall. In the case of vertical orebodies, pillar width is measured vertically; in a dipping orebody, the height of the underground pillar is equal to the stope width.

Unsupported span: the maximum unsupported length along the dip of the orebody. The estimated unsupported span takes into account the mining out of small pillars or their recovery to the extent that they do not provide the required support. Progressive collapse of pillars between levels means that the unsupported span can change with time.

Draw density: the assessed volume of extracted material divided by the extraction area. For simplicity, this is calculated as the depth of a sinkhole divided by a bulk factor for collapsed material. Draw density gives an indication of the self-supporting capabilities of the rock mass.

The mechanical properties of the rock mass include its strength (and the reduction of it over time caused by water and weathering) and low inter-block shear strength (provided by low joint roughness, and soft joint infillings and opened joints). The variability of all these factors results in random initiation and propagation of collapse leading to a sinkhole formation. A common cause of collapse is infiltration of water into the underground voids. Inflowing groundwater may cause rapid deterioration of mechanical properties of rocks due to its susceptibility to slaking or weathering. That is, it can soften the rock mass, reduce its bearing capacity and precipitate failure. All these factors can provide a trigger that initiates the collapse, which ultimately results in a sinkhole.

Review of nine case studies (Szwedzicki, 1992, 1999b) revealed that all collapses of crown pillar were preceded by precursors to collapse.

3.1.1 Coronation mine

Mining of the upper levels of the Coronation mine, located in the Eastern Goldfields, near Kalgoorlie, Western Australia, started in the late 1940s. Narrow veins of high-grade ore, dipping on average at 70°, were mined in irregular shrinkage stopes (Fig. 3.1). The width of the stopes was between 1.2 m and 1.7 m. The crown pillar was 10 m thick and consisted of oxidized rock. The pillars between the 1 and 2 levels and the 2 and 3 levels, located in an oxidized zone, were 1–2 m wide. These deteriorated with time and collapsed, leaving an open span (along the dip) of about 70 m. Below the 3 level, the pillar width increased to about 6 m.

In 1987, after heavy rain, a void progressed to the surface, forming a sinkhole with a diameter of more than 20 m and depth of about 12 m (Fig. 3.2). In time, the oxidized rock material around the wall fretted and partially filled the sinkhole. Analysis of underground maps indicated that the rock mass collapsed above 4 level, i.e. at 66 m below the surface, and progressed to the surface.

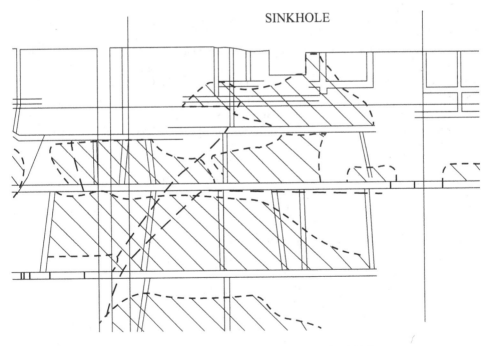

SINKHOLE

Figure 3.1 Longitudinal section through Coronation mine (Szwedzicki, 1999b).

Figure 3.2 Sinkhole above Coronation mine (Szwedzicki, 1999b).

3.1.2   Chaffers Shaft area

Uncontrolled ground movement resulting in the formation of a sinkhole and extensive surface cracks near Chaffers Shaft, Kalgoorlie, WA, took place in 1991. The movement caused
extensive damage to a road, which had to be relocated. At Chaffers Shaft, prior to 1960, ore
was extracted using shrinkage stoping, and all stopes in the area of the subsequent sinkhole
were dry filled. Underground observations showed that the dry fill did not transfer stress
from the hangingwall to the footwall and, when the supporting pillars failed, the fill migrated
downwards in steep stopes. The surface crown pillar was about 25 m thick and the resultant
sinkhole was 20 m in diameter and 8 m deep. The volume of the collapsed rock was estimated to be 3300 m³.

It was recorded that prior to sinkhole formation, the area was exposed to seismic
activity.

- twenty-seven years before, it was found that a seismic event damaged some pillars, and
  this allowed for dry fill migration to the lower stopes, and
- three months before the sinkhole developed another seismic event caused pillar failure
  and migration of dry fill into lower mined out areas. This seismic activity also caused
  localised failure of the back of the excavations. Such failure resulted in the creation of
  voids into which further rock could fall.

The sinkhole was immediately filled (Fig 3.3), but ground movement was observed to
continue for about two years resulting in subsidence of the fill material

Figure 3.3 Sinkhole in Chaffers Shaft area with fill material subsiding in time (Szwedzicki, 1999b).

3.1.3   Iron King mine

The Iron King mine, located in Norseman, Western Australia, operated during the Second World War. The massive, tabular pyrite orebody was mined by longhole blasting, which produced large and irregular unsupported stopes. The oxidized zone was 35–55 m thick and was formed of limonite and gossan. Above the mined-out area, a few sinkholes surfaced in early 1970. Two of them were of substantial diameter and depth.

One of these two sinkholes developed above the 5 level (about 80 m from surface). The orebody was 3–15 m thick, dipping at 50°. The unsupported span was about 55 m along the dip (Fig. 3.4). Records of the initial size of the surfaced void are unknown but some twenty year later the sinkhole was about 20 m deep and 40 m in diameter (Fig. 3.5).

The second sinkhole developed over a uniform pyritic orebody, which was about 10 in thick and dipped at about 45°. The lode was mined out from level 3 downwards to below level 5, i.e. from 50 m to 120 m below the surface. The unsupported open span was about 85 m and the crown pillar was about 50 m thick (Fig. 3.6). The sinkhole was elliptical with axes of 30 m and 45 m (Fig. 3.7) and when inspected some 20 years after its formation, it had a depth of 20 m.

3.1.4   Nobles Nob mine

The first ore mined from the 135 ft level at Nobles Nob mine, Tennant Creek, Northern Territory, was treated in early 1940. The surface collapsed over the mine in 1967. This resulted from the collapse of the surface pillar above stopes 4, 5 and 6, which had been extracted to

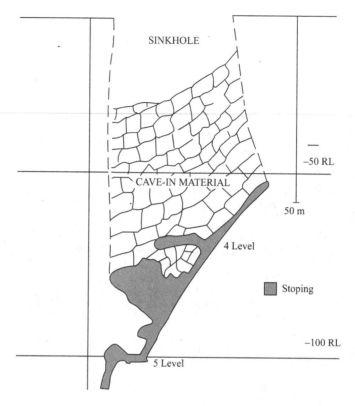

Figure 3.4 Iron King mine: longitudinal projection indicating sinkhole no. 1 (Szwedzicki, 1999b).

Figure 3.5 Iron King mine: sinkhole no. 1 (about 20 years after the subsidence).

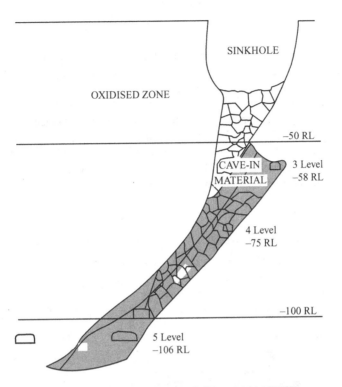

SINKHOLE

OXIDISED ZONE

−50 RL

CAVE-IN
MATERIAL

3 Level
−58 RL

4 Level
−75 RL

−100 RL

5 Level
−106 RL

Figure 3.6 Iron King mine, section through sinkhole no. 2 (Szwedzicki, 1999b).

Figure 3.7 Iron King mine: sinkhole no. 2 (about 20 years after the subsidence) (Szwedzicki, 1999b).

within approximately 25 m of the surface. The surface void was approximately 45–50 m in diameter (Fig. 3.8).

The main shaft winder, a Land Rover, a surface building within the area of the fall and 46 drums of cyanide were buried by the broken rock. The material which fell, estimated to be 80,000–90,000 tonnes, filled the underground workings in the area. The top of broken material was visible on the north side of the hole within 10 m of the surface. The sides of the hole were undercut to the south, from which a minor fall took place. The bottom of the sinkhole on the south side was about 22 m below the surface. Underground excavations were entered from another shaft to enable inspection on the 135 ft and 235 ft levels. It was found that no damage extended below the 135 ft level. It was observed that for the next three years circumferential cracks were opening around the sinkhole.

Figure 3.8 Crown pillar collapse at Noble Nob (1967) (Szwedzicki, 2001a).

Prior to the collapse ground movement in the vicinity of the underground void was recorded by an extensometer.

The sequence of precursors was as follows:

- About a week prior to the ground collapse, the extensometer in a drill hole recorded movement due to a large crack opening above the mined-out area.
- During the nightshift preceding the collapse, miners working near the area heard rock noises and reported four rockfalls.
- A major rockfall was reported one hour before the collapse.

3.1.5 Perseverance Shaft area

In 1958, a sinkhole developed on the Old Lake View Lode adjacent to the main shaft winder at Perseverance Shaft, Kalgoorlie. The area below the surface was honey combed with old workings. The stopes on the upper levels, many dating back to the 1890s, were partially filled. From an investigation of all accessible openings underground it was determined that the major subsidence was the result of ground movement at the 7 level and about 300,000 tonnes of ground subsided. The collapsing ground started to propagate to the surface from a depth of about 300 m. The sinkhole formed was some 35 m wide by 60 m long (Fig. 3.9)

Figure 3.9 Sinkhole behind a winder house at Perseverance mine (1958) (Szwedzicki, 1999b).

and about 23 m deep. Part of the winder room fell into the hole, and the north corner of the winder room was left hanging over the cavity. Since it was an extremely dangerous situation, the winder was dismantled and removed. The hole was filled with 50,000 m³ of waste rock dump.

The formation of the sinkhole was preceded by the following sequence of events:

- Sixteen weeks – an earth tremor took place that badly damaged one stope and 80 m of timbered drive.
- Fourteen weeks – a tremor took place that resulted in ground movement on the 15 and 17 levels.
- Five weeks – a mass blast before caused fracturing of rock mass in the vicinity of the stopes.
- Three weeks before, ground movement in the 1702 stope resulted in closure of the stope. The pillars on the 7 level failed and rock fragments ran into the 9 level. Consequently, the floor pillar below the 9 level was crushed.

It has to be noted that in the same location another sinkhole formed in 1942. There were no records of the event, but it was recalled that ground movement took place over a few days.

3.1.6   Prince of Wales mine

Operations at the Prince of Wales mine in the Eastern Goldfields, Western Australia, comprised underground workings and an open pit. Gold mineralization is hosted by a pillow basalt. Ore grade mineralization was associated with a near-vertical system of narrow carbonate veins. The upper levels of the underground Prince of Wales mine were mined out by shrinkage stoping, which started in 1947. A longitudinal section of the mine is shown in Figure 3.10. Open-pit mining took place in the 1980s; the depth of the pit was 28 m, its floor being only a few metres above underground stopes.

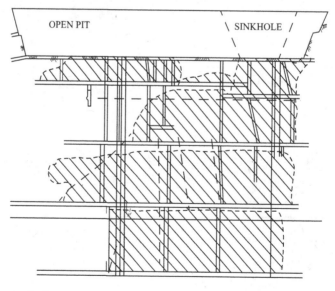

Figure 3.10 Longitudinal projection through Prince of Wales mine with superimposed sinkhole that developed in the wall of an open pit (Szwedzicki, 1999b).

A crown pillar failed during the 1991–92 New Year period. After heavy rains in December, a part of the bottom of the open pit and a part of a wall collapsed into two large underground open stopes (Fig. 3.11). Substantial failure of the pit wall resulted, which at the crest of the pit extended for more than 70 m. The ground movement exposed the shaft, located in the vicinity, and resulted in damage to the shaft lining. The thickness of collapsed rock was from about 10 m at the bottom of the pit to 38 m in the wall of the pit. The exposed open stopes were each about 2 m wide and 8–10 m long. The distance between the stopes was about 10 m. The caved material rilled to approximately the 5 level i.e. 122 m below surface. The equivalent diameter of the sinkhole was about 50 m. The collapse resulted in the closure of both the underground and open pit operations.

Figure 3.11 Prince of Wales mine: sinkhole through open pit and underground openings (Szwedzicki, 1999b).

3.1.7   Scotia mine

Scotia mine was located about 70 km north of Kalgoorlie, Western Australia. In 1974, a surface crown pillar collapsed above the stoping area. The area was in the vicinity of the shaft and the sinkhole engulfed part of the mine road. At the time of the subsidence the mine was operating on lower levels and, despite a large air blast, no additional damage was reported.

The orebody at Scotia dipped at 60–70°, flattening below the 830 ft (253 m) level. The stope hangingwall below the 230 ft (70 m) level consisted of serpentinite containing a large percentage of talc and chlorite. The surface crown pillar was approximately 60 m thick. The mining method was sublevel open stoping, and no pillars were left on the levels from 230 ft down to 660 ft. The upper stopes had been open for at least 10 years. At the time of the subsidence, the open span was 270 m and mining was in progress between the 660 ft and 830 ft levels. The first underground crown pillar was left on the 660 ft level and was 6 m wide. Robbing the pillar on the 735 ft sublevel resulted in stress transfer and deterioration of pillars on the 660 ft level. In 1974, a pillar on the 660 ft level collapsed, creating an air blast.

After the subsidence, the sinkhole was approximately 75 m long and 50 m wide, the collapsed material being approximately 30 m below surface at the northern end and rilling down to about 50 m below surface at the southern end.

Inspection underground revealed that the hangingwall had broken away immediately below the 230 ft level and the material had rilled down. The area where the hangingwall collapsed was known to comprise a poor, soft rock mass. Mining ceased because of the collapse.

The following sequence of precursors was noticed before the uncontrolled ground movement took place:

*   A few months before the main subsidence – it was noticed that the hangingwall above the 490 ft level was progressively deteriorating with large blocks falling into the stopes.
*   Three weeks – the floor pillar at the 660 ft level collapsed, resulting in ground movement across a shear zone on that level. The pillar collapse caused an air blast that lifted the concrete foundations of a surface fan.
*   A few days – progressive deterioration of the hangingwall and rockfalls were recorded.

3.1.8   Warrego mine

The Warrego mine, located west of Tennant Creek, Northern Territory, operated from 1967 to 1989. The main orebody plunged at 45° and dipped steeply to the east. The oxidized zone extended to a depth of approximately 100 m (in the range of 84–118 m). Mining was by longhole open stoping and left vertical pillars 20 m wide, transverse to the orebody. Primary stopes varied from 20 m to 65 m along the strike. The stope width between the hangingwall and the footwall varied up to about 70 m. Mining extended from the 2 level (60 m depth) to the 15 level (167 m depth). In the later stages of mining, these stopes were enlarged by mass blasting of the pillars. The strike length of the mined-out stopes was about 180 m, the plunge length 250 m and the thickness of the crown pillar was 160 m.

Significant ground instability commenced after a mass firing of a pillar, which took place some seven months before the collapse. At that time, a major hangingwall failure occurred over the middle stopes. About three months before the collapse, the surface subsided and a small surface depression was noticed. In the early hours of 4 June 1989, there was a collapse of the hangingwall, which progressed to the surface. The sinkhole thus formed was elliptical with a long axis of 40 m (Fig. 3.12), a minor axis of 20 m and a depth of 40 m.

Underground inspections revealed evidence of ground movement along the footwall contact and cracks were noticed in various places. The surface continued cracking around the sinkhole and ground movement was reported on the surface for the next few days. Cracking on the surface extended up to 80 m from the depression. About five weeks later, a second subsidence phase was noticed in the floor of the sinkhole. The third and fourth subsidence phases took place nine and twelve months later. The gradual ground movement continued until 1997, when the sinkhole was backfilled with waste rocks and tailings (Fig. 3.13).

Figure 3.12 Sinkhole, Warrego mine (1989).

Figure 3.13 Partially filled sinkhole, Warrego mine (1997).

The following is the sequence of precursors prior to collapse:

- Seven months before the collapse, mass firing of a pillar took place. This had instigated significant hangingwall instability, including a major hangingwall failure over the middle stope.
- During the period between the mass firing and collapse the following changes were noticed:

  - The shaft sump initially lost water but later a significant increase in water inflow was recorded.
  - Increased stress levels in the form of ground fretting and rock cracking were recorded around the shaft and on various levels.
  - Gradual movement was observed across the main fault.

- Twelve weeks – the surface subsided and a surface depression was noticed.
- Three weeks – airborne red dust was noticed settling on rocks in the 38 stope and along the 3-level access. (A layer of red dust was intersected on the upper levels when sinking the shaft.)
- Less than one week – a night shift miner working on the 3 level reported cracking noises and rock movement below the 3 level. An inspection by staff noted no obvious changes.

3.1.9   Analysis of surface crown pillar collapses

The mining depth from which the collapse propagated to the surface was obtained from the reports of underground inspections or estimated from the mining sections and plans. The deepest collapsed excavation that propagated to the surface was at a depth of 400 m. The distribution of the depth of cave-ins is < 100 m, two mines; 100–200 m, four mines; 200–300 m, two mines; and > 300 m, one mine. The volume of the material collapsed is a function of draw density (which ranged from 9 m³/m² to 46 m³/m²) and it ranged from 4000 to almost 180,000 m³ of rock material.

Summary of geometrical parameters and collapse time of sinkholes are given in Table 3.1.

To understand the behaviour of rock masses above mining openings, it is necessary to determine the extent of the open (unsupported) span. These were estimated as ranging from 70 m to

Table 3.1 Geometrical parameters and collapse time of sinkholes

Mine	Equivalent diameter, m	Thickness of crown pillar, m	Depth of cave-in, m	Unsupported span, m	Approximate collapse time, years	Volume of cave-in rock, 10³ m³	Draw density, m³/m²
Coronation	20	10	80	70	40	4.9	9
Chaffers Shaft	20	25	85	85	35	3.3	6
Iron King I	30	55	77	55	34	36.7	31
Iron King II	40	50	110	85	33	32.7	15
Nobles Nob	50	25	100	135	25	40.8	12
Perseverance	47	25	300	150	60	51.8	18
Prince of Wales	50	38	120	140	40	153.1	46
Scotia	63	60	265	270	10	182.3	35
Warrego	45	60	400	200	10	82.7	31

270 m. The estimates take into account that some small pillars were reduced in size to the extent that they did not provide the required support or that they would have collapsed. Figure 3.14 shows a relation between the diameter of the sinkhole and the estimated unsupported span. It indicates that sinkhole formation is sensitive to the geometry (length) of the unsupported span.

The mining excavations had very complex geometries because of the irregularity of the orebodies, varying stope shapes and pillar dimensions, and progressive pillar and hang-ingwall collapse. This made retrospective analysis more meaningful in two dimensions (sections) than in three dimensions (the full mine). An attempt was made to correlate vari-ous geometrical parameters i.e. thickness of the surface crown pillars, surface crown pillar width-to-thickness ratio, height and diameter of the sinkholes, volume of the collapsed mate-rial and ore draw density, and the stand-up time. The analysis did not allow the derivation of statistically justified relations between these geometrical parameters, except for a relation between the length of the unsupported span and the diameter.

In all cases there was a substantial interval between the mining and exposure of the sink-hole at the surface. Collapses occurred between 10 and 63 years after mining of the areas. In a few cases collapse took place after mine abandonment, whereas in other cases lower levels of the mine were being worked at the time of collapse. The mines that were still active closed their underground operations after the crown pillars collapsed.

Surface crown pillar collapse is a geotechnical phenomenon, which may be preceded by many years of deterioration in ground conditions and is followed by many years of post-failure ground movement.

In the case of the Perseverance Shaft, the first indication of possible collapse was sixteen years before the major collapse. That was the ground movement resulting in a void project-ing to the surface.

The following long-term indicators were recorded.

- In the case of the Perseverance Shaft, the first indication of possible collapse was 16 years before the major collapse.

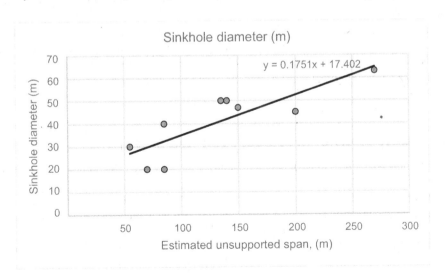

Figure 3.14 Relation between an unsupported span of the hangingwall and the sinkhole diameter (Szwedzicki, 1999b).

- In the case of the Chaffers Shaft: seven years before the collapse (during a seismic event), the damage to a pillar was recorded. There was a distinct pattern in rock mass behaviour during the period of a few months preceding the collapse.

  - Within 3–7 months before the failure, in four out of five analysed mines, the hangingwall started to deteriorate and progressive failure took place.
  - The collapse of the pillars between the levels and ground movement were recorded within one to three weeks.
  - Within days and especially in the final few hours before the collapse, rock noises and rockfalls were recorded. It is of interest to note that rockfalls were recorded in development drives and crosscuts some distance from the collapsing stopes. This is indicative of mining-induced stress changes taking place prior to the imminent collapse.

Analysis of the nine case studies shows that sinkhole formation in hard rock mining areas depends on many factors, and only very general relations were found between the geometric parameters of mining operations and the formation of sinkholes. This implies that risk analysis of sinkhole formation needs to be based on heuristic methods and should form part of the geotechnical risk management at any underground mine.

3.2   CASE STUDIES OF ROCKBURSTS AND OUTBURSTS

The intensity of rockburst and gas outbursts increases as underground mining goes deeper. Despite substantial research into mechanisms of these phenomena, it is generally accepted that prediction of such events in terms of magnitude and time has not been successful. The most frequently reported of rock mass response through precursors involves patterns of seismicity and acoustic emission. This has led to the hypothesis that a relationship exists between microseismic activity and rock mass damage (Gibowicz and Kijko, 1994).

Acoustic emissions are known to last for a few seconds. However, it was also reported that there was "progressive banging" lasting for some 15 minutes. Such progressive banging was reportedly accompanied by the liberation of large quantities of rock dust.

Four case studies on precursory behaviour of the rock mass prior to rockbursts and outbursts are presented.

3.2.1   Case study of precursors to rockburst

The Falconbridge Mine, Ontario, began operations in 1929. By 1984, about 94% of the ore had been removed. At that time, mining was at a depth of 1300 m and the method of mining was undercut-and-fill. Wide stopes were mined in panels with a maximum width of 5 m, and cut heights were typically 3 m. Seismic activity had been recorded since 1955. In 1981, a microseismic system was installed to provide an accurate record of seismic activity. The severity of the outbursts appeared to increase with depth, with the number of events reaching more than 60 per year in 1984 (Report of the Provincial Inquiry, 1986).

In 1984, a rockburst of a magnitude of 3.4 on the Richter scale shook the mine, causing a backfilled stope to collapse. The microseismic monitoring system located epicentres of the events along a series of faults in the immediate area.

The following were indicators of rockburst potential:

- previous history of seismic activity, and
- high extraction ratio.

The precursors that were recorded prior to the rockburst were as follows:

- Years prior the event – recorded seismic events caused no damage and resulted in displacements of less than one tonne of rock.
- Ten months before – a large seismic event displaced approximately 1500 tonnes of rock.
- Six days before – a seismic event with a magnitude of less than 1.0 on the Richter scale was recorded approximately 40 m from the stope in which the major rockburst occurred. No damage was found.
- During the six days immediately before – signs of excessive ground pressure (split timber and drift closure) were found.
- Five hours before – a seismic event was recorded in the vicinity, but it was apparently not large enough to warrant being reported to the mine supervisor.

A large number of research studies have been conducted on precursors and prediction of seismic events. Spottiswood (2000) states that short-term prediction of large seismic events does not seem to be possible. Glazer (2016) provided a review of research on precursors to seismic activities. The review concludes that 78% of seismic events occur directly after the blasting.

3.2.2   Case studies of precursors to gas and rock outbursts

Extraction of the coal seam in Nowa Ruda Mine, Poland, was carried out using the longwall mining method with caving. The area was known for its high level of carbon dioxide emissions, and the hazard of gas outbursts was recognized. Ventilation was continuously monitored to give early warning of increased content of carbon dioxide. The gas outburst hazard was assessed by measuring overpressure of carbon dioxide in drill holes and measuring desorption. The area had a history of gas outbursts. After a gas outburst in 1976, it was decided that further coal extraction could only progress after destressing the main coal seam. The place of the outburst was geologically disturbed, and the destressing holes allowed for determining the existence of a faulted and folded zone.

A large gas and rock outburst took place in 1979 when 150 tonnes of coal and 32,000 m^3 of carbon dioxide were ejected (Piatek, 1980). The following were indicators of rockburst potential:

- previous history of outburst activities,
- lower mechanical properties of coal in the disturbance zone,
- high levels of carbon dioxide pressure, and
- existence of a geological fault.

A year before the event, a large outburst ejecting 1170 tonnes of coal and 120,000 m^3 of carbon dioxide took place. The outburst was triggered by a central blasting. The following were recorded geotechnical precursors to that outburst:

- Six months – there was an outburst with 1500 tonnes of coal and 20,000 m^3 of carbon dioxide being ejected.

- Three months – signs of high stress were noticed (such as blowouts of gases and drilling cuttings) during drilling of destressing holes.
- Eight days – gas pressure in the drill holes could not be measured because the drill holes were closing.
- One day – the measured overpressure of carbon dioxide and desorption was above the critical level.

Possible triggers:

- Weakening of the rock mass by drilling exploration holes to determine geological disturbances.
- Water on the plane of a geological fault.
- Mining activities like drilling blasting holes, barring down of the face or using a mining equipment.

3.2.3   Case study of gas outburst

The Collinsville coal field in Queensland, Australia, was first developed in 1919. In 1954, a coal and gas outburst occurred at the No. 1 State Mine. It was estimated that 14,000 m³ of gas was emitted and 900 tonnes of coal and stone were ejected during the outburst. The coal and stone were blown 30 m up the heading. The outburst that expelled carbon dioxide was associated with geological faults. Damage after the outburst is shown in a section and in a plan view in Figure 3.15. The investigations by Biggam *et al.* (1980) and Sheehy (1956) revealed

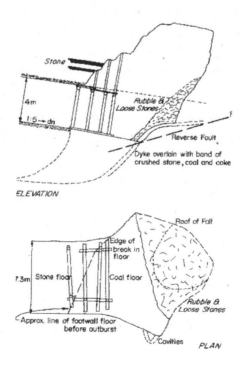

Figure 3.15 Outburst cavity (Biggam et al., 1980).

that the carbon dioxide was not uncommon in the workings and was often seen bubbling out of the water accumulated on the floor.

Indicators of outburst potential

- Previous history of emission of gas.
- Structural features like faults and fractures, dyke and igneous intrusions, seam disturbance and pulverized coal.

Precursors to the outbursts

- It was noticed that during two to three weeks prior to the outburst, the bubbling of carbon dioxide through the water accumulated on the floor had increased.
- On the day shift prior to the outburst, a disturbance in the floor was encountered. This consisted of broken soft coal and faulted coal showing slikenside lineations. This was recognized to be a sign of a fault that was evidenced elsewhere in the mine.
- The three men who were presumed to be working at the face must had have some warning, as their bodies were found well away from the face and uninjured by the flying coal.

Possible trigger

- Firing a small shot.

3.2.4   Case study of geothermal outbursts

The Lihir gold mine was located on Lihir Island in Papua New Guinea. Lihir Island is a volcanic sea mount rising above sea level. The orebody is contained within a caldera, and the Lihir mine hosts an active geothermal system which adds complexity and difficulty to the mine operations because of geothermal hazards (Villafuerte *et al.*, 2007). The deposit is hosted within extensively altered volcanic rocks, intrusive rocks and breccias.

Surface geothermal hazards manifestation in the mine area consisted of acid-sulphate hot springs, boiling mud pools and fumaroles. Mining operations exposed subsurface thermal manifestations and large areas featuring intensely steaming ground, geysers, boiling water and boiling mud. Hydrothermal alteration is also associated with unstable ground and landslips on site. Another geothermally derived hazard is presence of noxious gases, mainly hydrogen sulphide and carbon dioxide.

The major hazard to mining is the occurrence of hydrothermal eruptions, referred to as geothermal outbursts, which took place with varying intensities (Fig. 3.16).

During one of these outbursts, a shovel was free digging through argillic rocks overlying a dome of higher permeability. The argillic rocks served as an impermeable seal to the highly pressurized gas and steam that accumulated on top of the dome. Removal of the argillic rock cover reduced the lithostatic pressure that held down the gas/steam pressure in the underlying boiling zone, thus triggering the outburst. The trapped steam instantaneously depressurized in an explosive event.

Figure 3.16 Intensely steaming ground at Lihir mine.

A common feature emerging from the analysis of events leading to outbursts is that these events had precursors such as intensely steaming ground, local small outbursts and localized zones of high temperature and that they were potentially foreseeable.

Geothermal outbursts were controlled by cooling and depressurizing areas that were considered likely to result in an outburst when mined i.e. when a temperature probe measured more than 115 °C at the depth of 15 m or more than 130 °C at depth of 24 m. The methods of cooling and depressurizing included:

- drilling shallow steam relief holes,
- injecting water, and
- drilling and blasting in potentially outbursting areas and in adjacent rock mass.

3.3   CASE STUDIES OF UNCONTROLLED CAVING AND PILLAR COLLAPSES

It is difficult to find evidence of common rock mass behaviour prior to failure because literature on that subject is limited and geotechnical reports on rock mass failure are often confidential and not widely circulated. It is also difficult find common evidence, since many collapses happened after mines were abandoned and no records are available. Six case studies on large-scale ground collapses are presented, three from block caving mines and three from room-and-pillar mines. Case studies from block caving mines include investigation on propagation of the caving zone, caving progression to the surface and collapse of a caving area. Case studies from room-and-pillar mines include investigations into mass collapse of pillars in hard rock, coal and trona mines.

3.3.1   Discontinuous subsidence over a caving mine

At San Manuel Copper Mine, Arizona, USA, in 1956, block caving operations begun to mine 100 million tonnes of low-grade copper ore. Three massive porphyry orebodies were extracted at a depth of 160 m. The overburden consisted of conglomerate intersected by three major faults. It took 24 months for the ground to collapse after being undercut and ore draw commenced (Hathaway, 1968).

The surface above the extraction areas was monitored for any effect of caving. The process of ground caving was marked by two stages: a preliminary one of tensional fracturing together with gentle settlement of the ground surface followed by the final stage of sinkhole development. The first noticeable fractures at the ground surface were observed in a form of crater-like depressions, or "pits", that were directly above the caved areas. Contemporaneous with the formation of depressions, the surface tension fracturing developed in series of straight or arcuate fractures. With time, some these fractures extended to over 100 m.

For one orebody, fracturing was noticed on the surface about three months after the draw was initiated. The total time lapse of 560 days occurred between initiation of the draw and measurable ground settlement. However, once it started, the subsidence was rapid and large amount of rocks caved in.

Two years after mining began, subsidence was about 50 mm per day of vertical movement over the centre of mining activities and the first well-developed boundary escarpment began to form along the former fractures. As mining continued, the pit growth rate was quite uniform.

The surface area affected by subsidence progressively enlarged and stabilized after a period of seven years, with the final surface expression of the undercut being 800 m × 1000 m.

All rock mass responses and movement over the subsiding areas tend to channel material toward the pipe (sinkhole) areas. Based on the developed surface features, the following modes of rock mass response to caving were distinguished:

*   "En masse movement" that took place over the whole area that could be affected. En masse subsidence consists of initial vertical settlement of the area resulting in lowering of the rock mass. En masse subsidence was a precursor for other types of rock mass responses over caving areas.
*   Slump or lateral movement initiated by en masse deformation, which breaks rock mass into smaller blocks that moved into the centre of subsidence. Slump resulted in rotational movement and was generally slow.
*   "Terracettes" were a result of subsidence that forms terraces in the subsided areas. Usually terraces follow the outline of peripheral tension fractures. The terraces form as irregular step-like surface sloping towards the centre of the affected area.
*   Rockfalls and rock slides took place directly from the formed steep part of subsidence. They took place where subsidence lowered the surface so that dominant scarps were formed.
*   Creep and debris slide took place where material enters the pipes. Creep describes slow movement, while slide is often rapid.
*   Pipes, also called sinkholes, were a direct connection between the surface and caved-in areas. Pipes were centrally located over the active blocks in the mine.

The process of discontinuous subsidence over caving mines is similar to the formation of sinkholes over mining voids created by other mining methods. However, the major difference is in the scale of subsidence. In a case of subsidence over other mining methods, the diameter and the depth of the sinkhole are in the range of metres. In the case of caving mines, the diameter and the depth are in the range of hundreds of metres.

3.3.2 Propagation of a caving zone

PT Freeport Indonesia block caving mines are located within one geological complex and divided into three vertically stacked orebodies, which have been mined by Gunung Bijih Timur (GBT) Mine, Intermediate Ore Zone (IOZ) Mine and Deep Ore Zone (DOZ) Mine. Four years after production started, the area of the DOZ undercut was 79,300 m^2 and the perimeter of the undercut was 1.6 km. The designed cave over the DOZ Mine was over 600 m high, with total cave height from the production level to the surface being more than 1200 m (Barber *et al.*, 2000). The shape of the DOZ Mine caving zone was ellipsoidal with the short axis being 200 m and 400 m (width and length of the undercut) and the long axis being 650 m (height of the cave zone). The DOZ cave zone merged with the GBT and IOZ caves and reflected on the surface four years after production from DOZ commenced. Caving reflection on the surface, known as subsidence area (over three caving mines), was more than 1 km wide and 1.2 km long, with the total area estimated to be 1.2 km^2. The area of caving influence, i.e. within a crack line limit, was 1.7 km^2 (Fig. 3.17).

The shape of the caving zone was crucial from a draw control point of view. The caving rate and rock mass fracturing was important to determine the effect of caving on surface

Figure 3.17 Subsidence area over the block caving of two superimposed mines (Szwedzicki *et al.*, 2004).

facilities such as main exhaust fans and access roads. To monitor caving and subsidence, time domain reflectometry (TDR) cables were used to:

- define a draw rate,
- determine an air gap between broken ore and the crown pillar above, and
- determine the shape of the caving zone, caving rate and the effect of caving on rock mass fracturing.

TDR cables were installed on the DOZ undercut level (1200 m below the surface), the IOZ production level (900 m below the surface) and the surface (Rachmad and Sulaeman, 2002). The total length of monitored cables was over 14,000 m, with two of these TDR cables at 750 m long.

Results obtained for a period of over two years were analysed, i.e. from the first detected TDR breaks around the DOZ Mine to the cave break through to the surface. Results of TDRs breaks were calculated as absolute length of cables and their horizontal and vertical components. The vertical component of the broken length of cables was used to determine propagation of the cave to the surface. The horizontal component was used to determine lateral cave propagation, and the absolute length of the breaks was used to determine the extent of individual collapses of ground (Szwedzicki et al., 2004).

For vertical progression of the cave, the absolute length of broken-off cables varied from 10 m to 120 m. Out of 21 measured cable breaks, four were longer that 60 m, and it could be assumed that these breaks represented rock mass fracturing, i.e. formation of a dilation zone. Four cables broke at the length of 25 m to 40 m. The remaining 13 breaks were less than 20 m. Figure 3.18 indicates that the caving process was cyclic. During caving cycles, the TDR cables were broken in series over a period of two to four months. The periods of caving activities were intertwined by period of stability, lasting from a month to over two months.

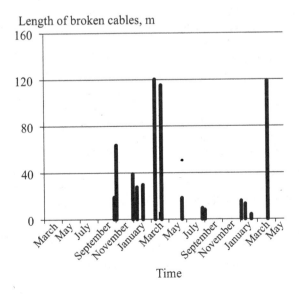

Figure 3.18 Periodical break of the TDR cables positioned over the progressive cave zone (Szwedzicki *et al.*, 2004).

Length of broken cables, m

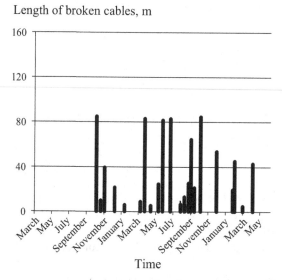

Time

Figure 3.19 Periodical break of the TDR cables monitoring cave horizontal progression of the cave line (Szwedzicki *et al.*, 2004).

It can be concluded that it was not the production rate which controlled cave progression but rather the total volume of removed rock.

To monitor the horizontal cave advancement in the southeast direction, thirteen TDRs were installed. In total, TDR cables broke 28 times because of the approaching cave. Figure 3.19 shows the absolute length of broken cables. There were 15 breaks shorter than 30 m, four between 30 m and 60 m, and seven longer than 60 m. The average length of the broken cables was about 36 m.

TDRs' breakage also progressed cyclically. The cycles lasted from a day to about a month, with periods between the breakages being about three weeks.

Analysis indicated that some of the TDR cables installed at the sides of the DOZ Mine broke in front of the undercut line and some behind. Breaking in front of the undercut line indicated rock mass fracturing in advance of the created void. As the result of that fracturing, a dilation zone was formed. Breaking behind indicated rock mass disintegration by caving.

Rock mass fracturing started immediately over the undercutting front and advanced in front up to 50 m. After the undercut lines reached their limits and advance stopped, some stress adjustments in the rock mass still took place, with the formation of new fractures continuing for up to six months.

In a number of cases, a number of TDR cables were broken simultaneously. With a wide coverage of the cables and large length of breakages, it can be assumed that the volume of cyclically fractured rock mass might have exceed millions of tonnes.

Cave growth or cave rate dictates the production rate. When caving slows or stops, the production must slow down or stop. It was found that the average cave rate was related to the class of rock mass (MRMR) (Szwedzicki *et al.*, 2004):

*   for poor rock mass, the cave rate was from 0.25 m to 1.10 m per day,
*   for good rock mass, the cave rate was from 0.15 m to 0.95 m per day, and
*   for very good rock mass, the cave rate was from 0.08 m to 0.30 m per day.

For a period of three years, starting from the first cave after reaching the hydraulic radius, the cave propagated vertically for estimated 645 m, i.e. the average cave rate was 0.6 m per day.

3.3.3 Collapse of the rock mass over caving area

The Northparks mine was using a block caving mining method to extract a copper-gold ore-body. The orebody was nearly vertical, of cylindrical shape and approximately 200–250 m in diameter. The plan was to extract it in two lifts – Lift One commencing at 480 m deep and Lift Two which was being developed at 800 m. Some challenges with caving were encountered early on. Caving started before the full footprint of the orebody was established. The caving resulted in an arched profile above the undercut, rather than the flatter one which was preferred (Bailey, 2003).

Because of the cave propagation problems, the air gap between the top of the muckpile and cave back progressively increased, as ore was drawn out. The air gap between the back of the cave and the muckpile increased to about 180 m because the production rate was greater than the rate at which the ore was falling from the back. To promote caving, a blasting campaign was undertaken with limited success, followed by a hydrofracturing campaign. The hydrofracturing campaign resulted in 2.5–3 million tonnes of induced caving. Although the method was successful in inducing the cave, it was unable to stimulate continuous caving. Caving activity was increased in the week or two before uncontrolled mass movement. As result, caving was starting to occur at an increasing rate. Coincident with that, the top of the cave passed vertically through the gypsum line, the line at which ore changed its characteristics, resulting in weaker material (Hebblewhite, 2003).

Rock mass response to caving and subsequently during hydrofracturing was changing and contributed to the uncontrolled mass movement. The following were precursors to uncontrolled mass movement:

- Two years to three months before caving difficulties in caving were reported.
- Several days – hydraulic fracturing and the drill and blast program led to significant propagation of the caving.
- Three days prior to collapse larger falls from the cave back occurred; this was estimated at up to half a million tonnes. Propagation of the cave back continued until the collapse.
- Fourteen hours – increased cave activity; large falls from the back.
- Two hours – caving rapidly increasing.
- Four minutes – 14.5 M tonnes fell in the stope.

The collapse was most likely initiated by progressive, localised caving (chimneying) from the hydraulic fracturing region up to the base of the pit. Further collapses, involving large volumes of material, were thought to have occurred in successive sequence, moving from the "chimney" outward and finally culminating in massive slope failure (Ross and van As, 2005).

3.3.4 Collapse of a roof due to pillar failure

Otjihase Mine, near Windhoek, Namibia, was extracting ore from a gently inclined copper orebody using room-and-pillar methods with the pillar factor of safety exceeding 1.5.

The ore zone, which plunged down, at 6 degrees, to more than 6 km long, consisted of A and B Horizons. Horizon A was 5 m thick and Horizon B was 6 m thick. The ore was mined using primary and secondary extraction. The primary extraction had 10.6 m wide stopes and 10 m wide pillars, making the extraction ratio of 51.5%. During the secondary extraction, pillars were reduced to 7.5 m, making stopes 13.1 m wide. The extraction ratio increased to 64%.

In 1987, the mining area began collapsing. The collapse started near the centre, inducing falls lower down and higher up in the mining area. The area was progressively collapsing, and in 10 days the collapse engulfed the decline. The collapse continued for over two years and covered approximately 175,000 m² (Klokow, 1992). Verbal reports from mining personnel verified that roof failure had been expected.

It was evident that months prior to the collapse, the following precursors were visible:

* roof spalling on the up-dip side of the pillars, and
* floor heave on the down-dip side of the pillars.

It was concluded that strata failure was likely to occur when secondary mining took place and the ultimate average stress exceeded the acceptable stress level. Two mechanisms could have initiated the collapse of the roof: delamination of the rock mass and shear on the up-dip side of the pillars in the roof and on the downside of the pillars in the floor, since in inclined workings the pillars and surrounding strata were subject to shear stress.

3.3.5   Collapse of pillars in a colliery

The disaster at Coalbrook North Colliery in Orange Free State, South Africa, took place in 1960 and arose from the collapse thousands of coal pillars over an area of at least 3 km² of the rock mass between workings in the No. 2 Seam and the surface (Bryan *et al.*, 1964). The seam was 8 m thick lying at an average depth of 150 m below the surface. The method of working was a board and pillar panel being ordinarily 240 m wide. Within each panel, 13 m square pillars were formed with drives of 2.5–3.5 m high. To increase coal recovery, secondary extraction was done by taking an additional 1.2 m of coal from the roof. This "top coaling" increased the height of working in the area to 3.6 m and, in places, 5.4 m. In 1957, the inspector of mines gave instructions for the height of working to be limited to 4.2 m. In 1958, "pillar cutting" operations, i.e. reducing the size of pillars, started. In the central part of the mine, the extraction ratio amounted to 77% to 83% (Moerdyk, 1965). The first collapse, which took place on 28 December 1959, was the start to progressive deterioration resulting in the second collapse on 21 Janaury 1960. It was followed by the third collapse three hours later, and then died away in the course of the next few days.

The recorded rock mass response to mining and the sequence of geotechnical events leading to collapse were as follows:

* Eighteen months – in a pillar, which had received four cuts, signs of crush were noticed.
* Thirty-three days – a major collapse of strata occurred in the upper two panels (out of three). During the next three days, rock noises at the perimeter of the fall persisted but then died down.
* Twelve days – a crack on the surface was noticed. It would appear that the barrier pillars at the perimeter of the fall, serving temporarily to arrest movement of the strata, were slowly failing.
* Three and half hours – miners heard sounds resembling shots firing and noticed that the sides of the pillars were splitting and coal falling. This was followed by sounds of heavy movement in the roof.

- Three hours – a miner in a section noticed a strong air blast. There was a sound resembling heavy thunder. The mine overseer found that two of the stopings had been blown down and that firedamp was issuing. Cracking noises were noticed.
- Three hours and again one hour – the formation of cracks on the surface was observed. Large cracks appeared in the main road and a considerable depression formed. The cracking on the surface progressed for a distance of 1.3 km.
- Twenty minutes – thundering become more frequent and the quantity of firedamp issuing increased.

Upon collapse, the thunder sound of collapse was accompanied by a hurricane of dust-laden air. A great part of the mine had fallen in, the collapse released a great quantity of methane and the gale blew with great velocity for three-quarters of an hour.

Following the disaster, a circular depression 2 m to 2.5 m deep had formed above two panels. The area of subsidence was at least 3 km^2. Strata movement continued for many days. A survey showed that over most of the affected area, the surface subsided by about 0.6 m.

Three tremors, which were identified as originating near the Coalbrook Colliery, were recorded by four seismological observatories. The tremors 24 days before and 2.5 hours before were single shocks with duration of a second or two. The relative severity at the epicentre was between 0.3 and 0.5 on the Richter scale. Seismic records show that there were small vibrations about 3.5 hours and 2.5 hours before the event. The tremor at the time of collapse continued for five minutes with three separate peaks of amplitude. The severity at epicentre was 1. Slight vibrations occurred for a few days after the collapse.

3.3.6   Progressive collapse in room-and-pillar trona mine

A large-scale collapse occurred in a room-and-pillar trona mine in 1995. An area measuring approximately 1 km by 2.5 km collapsed. The Solvey Minerals mine, Wyoming, produced trona from a deposit at depth from 460 m to 520 m. In the collapse area, continuous-bore mines were used for production. The room width was 4.6 m, chain pillars were 12 m by 14 m and panel pillars were 3.8 m (Amadei *et al.*, 1999).

Miners working near the panel that collapsed reported that they heard a rumbling, a big boom and then a deafening sound lasting five to six seconds. The collapse resulted in damage to the mine, an air blast and ammonia and methane emissions (Goodspeed *et al.*, 1995).

Panels in the southwest section of the mine collapsed completely or were severely damaged. There was no collapse of panels elsewhere in the mine. The collapse involved both of the pillars and the immediate floor material. The upper part of the pillars and immediate roof remained intact, and the failed rock erupted from the immediate floor and lower pillars. Cutter-type roof failure was generally observed in some gateroads, but not in production rooms. The entire surface area over the panels subsided between 0.75 m and 0.9 m. Prior to collapse, little subsidence was observed over the southwest section.

An estimated 2.8 million m^3 of methane were liberated from the collapsed area, presumably from shales in the roof and floor. Methane emissions from the mine returned to normal levels of about 28,000 m^3/d about three months after the event.

The collapse event registered 5.1 M$_l$ (local magnitude) and was one of the largest events associated with mine collapse. It was concluded that the seismic energy emanating from the vicinity of the mine was not tectonic or natural earthquake but rather the recorded seismic energy resulting from the mine collapse.

It was suggested that the mine collapse was a "progressive pillar collapse", also known as "domino-type failure". In this mechanism, when one pillar fails, the load it carried is transferred rapidly to adjacent pillars, causing them to fail and leading to rapid collapse of very large areas of a mine. Once it begins, it is self-propagating. Collapse continues until all pillars fail or when a solid abutment is reached.

It was proposed that a triggering mechanism was the gradual degradation of strength in the panel pillars. Production rooms throughout the mine were associated with floor heave and pillar degradation that increased with time following mining. It was concluded that local loss of pillar-floor load-bearing capacities, with further reduced mine stiffness in the immediate surroundings, initiated the unstable progression of pillar-floor failure (Zipf and Swanson, 1999).

3.4 CASE STUDIES OF DAMAGE TO UNDERGROUND MINING INFRASTRUCTURE

Mining infrastructure like orepasses, crusher chambers, magazines, declines and even shafts are occasionally situated relatively close to ore extraction zones. They often are subjected to mining-induced stress and seismicity, resulting in ground deterioration, fall of ground or rock mass fracturing, which can ultimately lead to collapse. The risk is even higher when the infrastructures are intersected by geological structures or are located in poor-quality rock mass.

3.4.1 Enlargement of orepasses

Orepass operability is a key performance indicator for many underground mines. Orepass conditions depend not only on mining-induced stress and rock mass properties but also on the properties and volume of the transferred material and, arguably, the ground support installed in the passes. Material travelling down a pass causes attrition through damage by the impact of falling and bouncing blocks (McCarthy and Askew, 1986). Reviews of orepass design (Hadjigeorgiou et al., 2004; Stacey, 2004) indicate that ground support in the form of bolts do not necessarily increase orepass wall integrity. It is argued that falling material impacting on bolts can cause steel vibrations that result in rock mass degradation and deterioration of the orepass walls.

A case study is provided on orepass deterioration, i.e. increase in diameter. In a sub-level caving mine, three orepasses were raisebored 883 m below surface. The raisebored orepasses were vertical, 170 m high and 3.5 m in diameter. The orepasses extended over multiple levels and were fed through "tipples" (finger rises connections to the passes). The tipples were typically fitted with grizzlies and inclined at 70–80 degrees. All of these passes were in the same geological formation. The orepasses were supported at the brow on the bottom level with 2.4 m chemically grouted bolts and mesh.

The stability of orepasses were affected by abutment stress caused by the progressing extraction front. Seismic monitoring indicated that the most of seismic events were recorded about 50–100 m from the extraction front. With extraction progressing down dip, the abutment stress superimposed mining-induced stress round the orepasses. It could be deducted that instability in the orepasses was progressively moving down dip. It was realised that there was the potential for the orepasses to become inoperable. Should these orepasses be

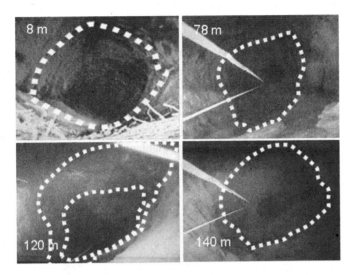

Figure 3.20 Change in shape of the orepass at 88 m, 78 m, 120 m and 140 m three years after commissioning.

unserviceable, without a replacement orepass system, the mine would need to convert to ore and waste trucking.

During raiseboring of the orepasses, stress- and geological structure-related damage was recorded for all three passes. Ten weeks after the first orepass was completed, it was surveyed using a video camera. This showed that there was no damage to the top 90 m of the pass – the raise was circular and as designed. From 90 m to 145 m from the top of the pass, stress (in the form of breakouts) and structure-induced damage was observed to be up to 0.6 m deep into the pass walls, creating a long axis dimension of 4.1 m. Below 145 m from the top of the pass, the raisebore was elliptical and further enlarged to almost 5.5 m.

About three years after raiseboring, the orepasses were surveyed using a video camera. This survey showed that the pass had suffered further substantial breakouts and enlargements, which were both structurally and stress controlled. Pictures were taken at depths of 8 m, 78 m, 120 m and 140 m below the collar of the orepass (Fig. 3.20). The broken lines in Figure 3.20 indicate the inferred profiles of these sections.

Four years after raiseboring, a survey of the orepasses was carried out using a laser scanner that was capable of collecting images of walls, from its sideways-looking cameras, and to determine multiple orepass cross-sections (Jarosz and Langdon, 2007). At the time of inspection, the orepasses were half-full and only the upper part was scanned. By the time of that survey, the total combined material throughput was 295,000 tonnes through the first orepass, 330,000 tonnes through the second orepass and 135,000 tonnes through the third orepass.

The 3-D solids, constructed from the laser scans (Fig. 3.21), showed substantial deterioration in the ground conditions around all of the passes. Noticeably, the scan results showed that the two passes had enlarged by up to 20 m at a distance of 75 m below their collars. The deterioration appeared to be exacerbated by a geological structure running north to south (parallel to the orebody) and intersecting these two passes (Fig. 3.21). The main direction of enlargement was in the north–south direction, subparallel with foliation and perpendicular to the maximum principal stress direction. At 100 m below the orepass collar, the limit of the survey, the third pass had broken through to the northeast corner of an access drive. Also, the orepass had enlarged by 5 m, making its maximum cross-section 8.5 m.

Analysis of horizontal slices through the passes (based on laser scanning) enabled the interpretation of the predominant modes of failure. From the collar of the pass to a depth of about 25 m, the orepasses exhibited stress breakouts perpendicular to the direction of the maximum principal stress. At depths of 30–35 m, where finger rises connecting to the level tipples were excavated, the predominant mode of failure was controlled by blast damage. From depths of 54 m, where geotechnical core logging had identified a fault, the predominant mode of failure was structurally controlled (Fig. 3.22).

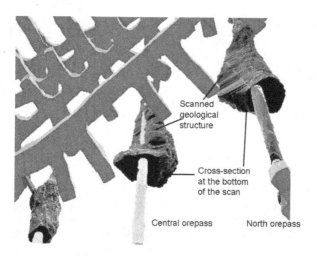

Figure 3.21 Bottom-up view of the passes, showing damage due to intersection of an inferred structural plane.

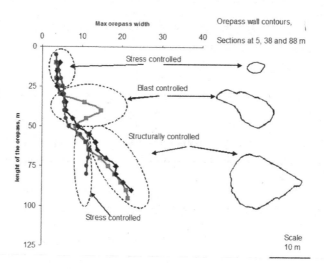

Figure 3.22 Modes of failure of the orepasses.

Four years after orepasses commissioning, there was evidence that the orepasses, including tipples and level accesses, would not be sustainable for the revised and extended life of the mines. Deterioration in the ground conditions surrounding the orepasses was indicated by:

- substantial enlargement of two orepasses, with a relatively smaller enlargement of the third one,
- higher than expected pass attrition and occasional slabs (assumed to be from the pass walls) reporting to the bottom of the passes, and
- ongoing seismicity in the vicinity of the lower levels of the orepasses.

A few hang-ups took place in the initial phase of material transfer. However, with the cross-section enlargement only large blocks caused blockages at the bottom of the passes.

In properly designed and maintained orepasses, there are two main reasons for hang-up formation (Szwedzicki, 2007a):

- numerous large blocks (oversized rocks) fed into an orepass at the same time can result in interlocking and wedged blocks, and
- wet fines ("sticky mud") can form compacted arches when draw from an orepass is intermittent (even for a single shift).

From a practical point of view, an orepass fails when its diameter is substantially enlarged, which typically results in huge blocks detaching from the wall (Fig. 3.23). The blocks not

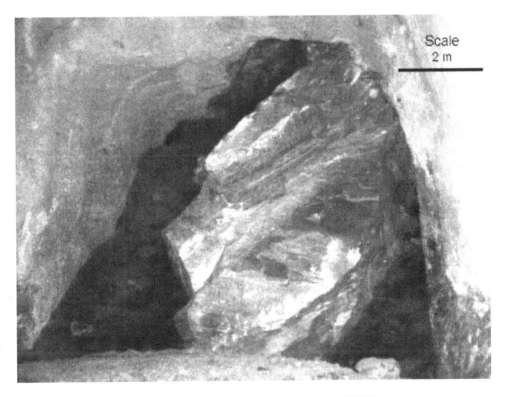

Figure 3.23 Blockage to a drawpoint due to wall collapse (Szwedzicki 2007a).

only cause dilution, but more importantly they also block drawpoints. Critical failure is attained when an orepass is large enough that it affects the stability of other excavations adjacent to it.

To prevent the formation of hang-ups, the following operational practices were implemented:

- Continuous drawing of material from the passes. Material within the orepass that contains a high content of fines should be constantly moving. An interval between the consecutive draws should be as short as possible and should not exceed eight hours. Passes must be regularly operated to keep the material moving.
- Breaking of oversized blocks on grizzlies. A large number of oversized rocks should not be passed through a grizzly at the same time (e.g. end of a shift). Despite breaking by a rock breaker, a number of large-sized blocks sent through the grizzly at the same time tended to form hang-ups. Oversized rocks should be gradually sent through a grizzly, with no more than three blocks sent down the grizzly before the introduction of more fragmented rock.

The designed volume of each raisebored pass was about 2000 m³. At the time of the survey, i.e. four years after raiseboring, the volume of the first orepass was estimated to be 12,000 m³, the second orepass 10,000 m³ and the third orepass 6000 m³. This meant that the volume of the two orepasses increased five to six times, while the volume of the third orepass increased three times, which indicated that the orepasses were liable to become inoperable.

Orepass inoperability could result from:

- damage to infrastructure (tipples, drives, etc.), or
- hazards from the collapse of other openings in the vicinity.

Orepass integrity was managed by implementing strict operating procedures to choke feed the system and control material size. To provide lateral confinement and thus prevent sloughing of loose material from the wall of the passes, it was imperative that a "choke-fed" state for the passes was maintained. The material level in the orepasses needed to always be kept as high as practicably possible. The minimum material level was determined to be 82 m below the top of the pass. This allowed tipping into the passes from three levels while still providing some surge capacity within the pass system. With the passes choke fed to the defined limit, the surge capacity was limited to about 1000 tonnes (in each pass).

To determine whether an orepass had become inoperable and should not be used any more, the following criteria for decommissioning of the passes were derived:

- Inoperable – excessive wear resulting in breakthrough to other excavations.
- Unserviceable – no safe access.
- Collapsed – progressive failure, permanent blockage.
- Enlarged to seven times the original diameter, (i.e. 7 m × 3.5 m ~ 25 m).
- When rock blocks detach within an orepass above a tipping level.

Analysis of the pass behaviour allowed for the formulation of design principles for an orepass replacement strategy. In expectation of inoperability, orepass closure criteria were derived and an orepass replacement program was proposed.

Closure of a crusher chamber

The first warning signs of ground deterioration around a crusher chamber were noticed in a form of convergence. This was followed by damage to support, fall of ground, acoustic emissions and ground deterioration in the crusher and its accesses. As a result, based on risk analysis, a decision was made to decommission the crusher by ceasing production and initiating salvage of the crushing equipment. Increasing the scale of geotechnical events resulted in suspension of salvage but subsequently, as geotechnical events subsided, a decision was made to recommence the salvage operation.

The crusher chamber was 12.5 m wide, 11 m long and 17 m high. During construction, the crusher was supported with friction stabilizers, chain link mesh, W-straps and cable bolts.

About three years before the closure, it was observed that ground conditions started to deteriorate slightly around the crusher chamber and sporadically in other excavations near the crusher. After horizontal convergence was detected, additional cable bolts were installed and the whole chamber was shotcreted. The deterioration was caused by the progression of a caving zone of a neighbouring mine (about 350 m away).

Some of the major recorded and monitored events during the six months preceding the closure are described below (Szwedzicki and Sahupala, 2004):

- Damage to a horizontal beam installed across the crusher chamber. A steel beam stabilizing the crusher hopper and overhead crane started to buckle (Fig. 3.24), about three years before the crusher closure. After the initial deflection was noted, several convergence stations were installed. A year before closure, monitoring showed a steady increase in convergence reaching up to 0.30 mm/day. Total cumulative displacement measured for that year was about 0.25 m.

Figure 3.24 Bending of a steel beam due to horizontal closure (Szwedzicki *et al.*, 2004).

- Damage to steel arch support in a ventilation drift. Substantial deformation that led to the collapse of the steel arch support took place gradually over a period of almost 18 months prior to the crusher closure. The collapse took place where a fault zone intersected the drift. The volume of the fall of ground was in excess of 150 m³.
- Damage to a conveyor drift. About six months before closure, it was noticed that the ground started to deteriorate at an intersection between the conveyor drift and the crusher. The intersection was situated in marble, which was considered weaker than the surrounding rock mass. At the beginning, vertical cracks were noticed and, within a month, the rock mass between floor and grade line started to spall and move. Despite installation of friction stabilizers, mesh and W-straps support, about five months before the closure, a rockfall took place. The volume of the fall was about 7 m³.
- Ground deterioration on the extraction level. Ground and support deterioration was noticed on the extraction level, above the crusher. Around six months before the closure it was observed that the shotcrete was cracking and spalling off and the ground was unravelling. Just two weeks before closure, a fall of ground occurred immediately above the crusher and a few pillars in the area were yielding.
- Ground deterioration in the crusher. Support in the crusher showed signs of deterioration. About three months before the closure, shotcrete started to crack on both sides of the entry to the crusher. The cracks were vertical or almost vertical, indicating that the shotcrete failed in extension due to vertical principal stress. At the time of closure, the largest crack was about 3 m high and 150 mm wide
- Damage to the intersection of service and drainage drifts. The intersection between these two drifts was heavily supported with concrete walls and steel support. Two months before the closure, a crack was found in the concrete sidewall beside a steel set. A month before the closure, ground deterioration accelerated with new cracks developing at the shoulder height. Within a week prior the closure, welding on the steel caps were sheared and steel posts were moved from their place, pushed by ground behind the steel sets. Steel cap elements were bent and twisted.
- Rockfall at a feeder. Three months before the closure, a large fall of ground took place about 4 m away from the chute of a feeder. The highly fragmented ground (about 150 m³) buried a part of the tail of the conveyor. The rock mass consisted of weak marble and sandstone and was supported with friction stabilizers and wire mesh. Investigations after the event revealed that the support was heavily corroded. After the rockfall, water seepage was noticed through fractures. Geotechnical monitoring of convergence stations prior to the fall showed that convergence increased to 0.35 mm/day. After the rockfall, ground deterioration propagated along the conveyor drift and was progressively accelerating prior to the closure. Cracking, spalling and slabbing took place at the shoulders, ribs and backs along the drift. At the time of closure, water started to seep along the whole drift.
- Squeezing of drawpoints on the production level. A day before the closure of the crusher, drawpoints on the extraction level 30 m above the crusher quickly deteriorated and collapsed. Concrete-supported ribs and floors of two drawpoints moved inwards, resulting in abandonment of the area. Steel support at the brows of the drawpoints was damaged, and the rock mass was heavily fractured. The convergence movement was followed by inflow of water through cracks.
- Seismic events and acceleration in damage to the conveyor level. In the morning on the day of the crusher closure, a seismic event in a form of a large acoustic emission

was heard in the vicinity of the crusher. An immediate geotechnical inspection revealed quickly progressing damage to at least six areas on the conveyer level. The most evident were new cracks in the walls of the crusher, further ground deterioration along the conveyor drifts (fracturing, spalling, small falls of ground and convergence), and increased damage to shotcrete and steel support as well as to the structural elements of the conveyer system. In a few hours, additional damage to excavations and a rockfall were recorded.

To conduct salvage in safe working conditions, geotechnical monitoring was carried out twice a day with the results analysed providing immediate feedback to decommissioning crews. To monitor ground movement, a year before the closure, two convergence stations were installed across the crusher chamber, and six months before the closure, 12 stations were installed on the conveyor level. During deterioration in ground conditions (three months before closure), measurements were taken once a day and during the period of salvage even twice a day. After analysing changes in convergence, working permits were issued to salvaging crews. The permits were issued based on criteria established on the value of daily change in convergence (Rachmad and Widijanto, 2003):

- < 1 mm/day – no hazard of uncontrolled ground movement; work can continue,
- 1.0–2.0 mm/day – a limited hazard of uncontrolled ground movement, short time activities, under supervision, and
- > 2 mm/day – hazard of uncontrolled ground movement; work to be suspended until further notice.

For the period of three months before the closure, the average daily convergence was 0.45 mm. After suspension of salvage activities, the daily convergence varied from 0.21 mm to 1.68 mm with a trend line decreasing 0.6–0.15 mm/day.

Analysis of the sequence of the geotechnical events and subsequently conducted risk analysis resulted in a decision to suspend the salvage operation and to withdraw crews from the crusher.

The progressive damage to the crusher and its surrounding area, from the moment the first signs of deterioration were noticed to the moment the crushed operations were suspended, took about three years. Although there were local damages or even failures, there was no mine scale (crusher chamber scale) failure. For about one month, there was a period of acceleration in deterioration, which culminated in a number of seismic activities. Three large and a number of small seismic events were recorded in six hours. These events were felt for a short period of a few seconds. The last event lasted for about 15 minutes, with "progressive banging" heard in the roof of the crusher. The last event was accompanied by the liberation of large quantities of rock dust in the crusher. There were reports that earth tremors were felt at a distance of 5 km to 10 km away. An immediate geotechnical inspection and results of geotechnical instrumentation did not show any substantial changes in damage in the vicinity of the crusher.

After that culmination, a number of residual events were recorded, such as:

- Small rockfalls were recorded in an exhaust drift. The exhaust drift was situated 56 m below the crusher and 200 m to the west of the caving zone.
- Loose rocks detaching from the back were found in many places.
- A long fracture in the floor about 80 m long was noticed on the incline.
- There was a change in water inflow pattern.

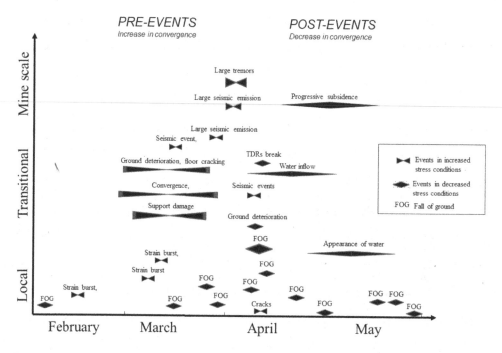

Figure 3.25 Types of ground deterioration before and after a major event.

After seismic activities subsided, daily changes in convergence were reduced to 0.8 mm or less. It was also noticed that ground deterioration ceased. After eight days of monitoring, it was concluded that mining-induced stress had redistributed and that it was safe to continue salvage operations. With time, the number of geotechnical events subsided, convergence reduced and there was very little progressive damage.

In retrospect, looking at the cumulative convergence, the decision to close the crusher operations was made when the convergence displayed a transgressive trend, i.e. daily changes in convergence were increasing (Sahupala and Szwedzicki, 2004). The decision to recommence salvage operations was made when the convergence started to display a regressive trend, i.e. daily changes in convergence were decreasing.

It is interesting to note that before and during the merge, deterioration/damage predominantly took place in increasing stress conditions such as seismic events, convergence or ground support failure. After the merge the deterioration/damage to the rock mass predominantly took place in decreased stress conditions such as rockfalls, water seepage/appearance, crack opening. Figure 3.25 summarises the geotechnical pre-events and post-events during the period of increased deterioration. The events depending on the severity and extend were classified on local, transitional and mine/regional scales.

Propagation of the zone of caving resulted in changes in ground conditions at the conveyor level. Early warning signs and the rock mass precursory behaviour indicated that rock mass was under high mining-induced stress.

Analysis of geotechnical events proved that:

• The first geotechnical warning signs leading to major ground deterioration in the crusher and the whole conveyor level were detected three years prior to the decision of closure.

- Geotechnical events such as ground deterioration, rockfalls, support damage, convergence and seismic events increased in time and severity. Large-scale ground instability was preceded by small-scale events.
- The recorded geotechnical events were noticed where the rock mass had the lowest mechanical properties, geological structures were present, mining excavations had large open spans and ground support was corroded.

Damage to the structure of the rock mass was followed by a change in water inflow pattern.

3.4.2   Ground movement in a decline

A case study is presented on the effect of stoping on stability of a decline sited in close proximity of a stope. The problem was compounded by the existence of a shear zone which was associated with the orebody and intersected the decline. The decline, situated in a footwall formation, was an important access to the lower levels and had to be maintained (Szwedzicki, 1989b).

It was expected that when stoping the orebody, the ground conditions would deteriorate and that a slip along the shear zone might have resulted in severe damage to the decline. The position of the decline in relation to the stope is illustrated in Figure 3.26. The footwall

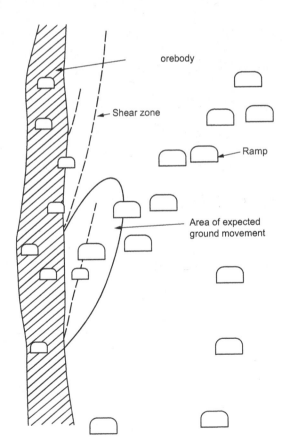

Figure 3.26 Section showing a shear zone intersecting a decline (Szwedzicki 1989a).

side of the zone was competent, and the hangingwall was blocky with distinctive joints and open cracks. The shear zone comprised at least three shear planes dipping from 70 to 90 degrees and were parallel to the orebody. The zone was filled with clay, talc and chlorite, and its thickness was about 30 m. Movement across the zone was recorded in the decline and it appeared to be linked to the failure of pillars.

The relative movement across the zone was monitored periodically by registering the movement across the paint marks. The observation marks showed that over eight years, the movement varied on different levels, ranging from 70 mm to 250 mm. Figure 3.27 shows flexure of hard rocks due to movement along a shear zone in high stress conditions. Fracturing and considerable vertical movement resulted from regional ground movement and from the fact that the decline had been undercut by a footwall drive. Vertical distance between the floor and the back of the drive in this area was only about 2.5 m. To monitor the effect of stope blasting on the movement across the shear zone, measuring plates were installed (Fig. 3.28). The plates showed steady movement at a rate of 5–7 mm per three months before stope blasting started. Blasting activities increased the velocity of movement, and the total movement in four months was 35 mm. When blasting was suspended, the measuring plates

Figure 3.27 Flexural deformation of the rock in a sidewall of a decline (Szwedzicki 1989a).

Figure 3.28 Measuring plates installed across the shear zone (Szwedzicki 1989a).

revealed no further displacement, signalling the attainment of a new state of equilibrium. Movement recommenced when blasting recommenced. The stope was successfully blasted and was filled with sand.

Had the fault and shear zone been mapped on the upper levels, the decline would have been situated further in the footwall.

3.4.3   Shaft collapse

A case study of the collapse of a shaft headgear due to formation of a cavity behind the shaft lining is briefly presented. A shaft of 8 m in diameter was sunk for a deep underground coal mine from the surface through 5 m of soils and into sandstone and mudstone sequences inclined at about 10 degrees. The uppermost sandstone stratum was 18 m thick, was characterized by low strength and was water bearing. Water drained through this sandstone and through cracks in the concrete lining to the shaft. The amount of water seepage was considered negligible and was not controlled, and no water testing for chemical content was carried out.

After 25 years of such conditions, the headgear collapsed. This was a gradual process: the duration from the first moment the movement was noticed until the time of collapse was 30 minutes. Subsequent investigation revealed that the water which drained into the shaft was slightly acidic and dissolved the sandstone bonding and washed out the fine sand, creating a large cavern. This cavern was located directly beneath the soil down to about 20 m below the surface. The headgear subsequently became unstable and collapsed. This serious failure could have been avoided if the water had been chemically tested to enable preventative action.

Chapter 4

Case studies of rock mass response to surface mining

Slope instability creates substantial hazards in open pits, tailings dams and mine waste dumps. In open pit mining, large slope failures are preceded by measurable displacement. Once displacement has been observed, 75% of slopes collapse. Of these, 30% fail within three months and 60% within two years (Sullivan, 1993). However, some mine slopes have been in an advanced stage of movement for decades. Regarding rock mass failure in open pits, Call (1992) noted that slope failures do not occur spontaneously. A rock mass does not move unless there is a change in forces acting on it. The common factors that lead to instability in an open pit are removal of ground that provide support to the walls, increased pore pressure and earthquakes. A slope failure does not occur without warning. Prior to major movement, measurable deformation and other observable phenomena occur from hours to years before major displacements (Szwedzicki, 2003b).

Three case studies of open pit slope were reviewed. In these studies, the geotechnical modes of failure and precursors to instability that were identified in hindsight are given below.

4.1 OPEN PIT SLOPE FAILURE DUE TO UNDERGROUND MINING

Sons of Gwalia Mine commenced in 1897 as an underground mine. It ended operations in 1963 and produced over 2.5 million ounces of gold. During that period, the orebodies were extensively mined out using the cut-and-fill method. Most of the mined-out stopes were filled with waste rock and tailings. The open pit mining commenced in 1984 and closed in 1999 (with a pit depth of 245 m). In 1999, underground mining recommenced using long-hole open stoping, with subsequent backfill.

In April 2000, the east wall of the open pit collapsed (by subsidence and toppling), and the rock mass surrounding underground excavations were subject to stress and ground movement. The collapse extended 250 m along the strike and was 200 m thick and 280 m high. It was estimated that about 40,000,000 tonnes of rock mass subsided.

From the first observation of the formation of tension cracks to the final collapse, the area was under detailed geotechnical monitoring (Varden, 2001). The damage caused by collapse was observed on the surface and in underground excavations. On the surface, the portal to the underground excavations was substantially deformed, the wall of the pit was cracked and cracks behind the crest of the wall extended to about 200 m behind the pit. The underground excavations were subject to rockfalls, floor heave, crushed pillars and extensive ground movement from 245 m to 315 m below the surface (down to 70 m below the pit bottom). The analysis indicated that the toe of the pit wall was pushed up and out into the pit. This movement was evident on the first underground mining level 10 m below the pit floor.

The sequence of events leading up to the collapse was as follows:

- Four months prior to collapse – tension cracks were noticed around the portal and through-out the mine and monitoring of surface prisms situated along the wall started to show sub-sidence, with the vertical movement increasing form 0.08 mm to 0.5 mm per day.
- Three months – further movement of prisms, stress signs in underground excavations in a form of shotcrete cracking, and ground deterioration and small rockfalls.
- Two months – a slight decrease in prism movement, ground movement around under-ground excavations, widening of existing cracks and small rockfalls.
- One month – number of small falls of ground and crack widening, rock noises, and cracks opening behind the crest of the east wall.
- Throughout out the week before the collapse, rock noises were heard in four locations. The frequency increased before the failure.
- One to four days – increase in number of rock noises, split sets failure in shoulders of the drives, four falls of ground and increased ground movement recorded by prisms.

Vertical movement of the prism installed in the east wall, recorded for four months prior to the collapse, totalled 180–200 mm. The average vertical movement of the wall during the collapse was in a range of 600–700 mm.

It was postulated that historical mine workings contributed to the rock mass failure, and that the failure was triggered by a heavy rainfall.

4.2   SLOPE FAILURE ALONG GEOTECHNICAL STRUCTURES

Gold mining activities in the Pine Creek area of the Northern Territory, Australia, began in 1873 when gold was discovered during construction of the Overland Telegraph Line. In 1875, the rush for gold saw an influx of miners to the Union Reefs mining area. This resulted in about 2000 small workings. Excavation of the Crosscourse open pit mine at the Union Reefs Mine began in 1994. The rock mass of the Union Reefs Mine area consists of weath-ered, closely bedded (0.10 m to 1 m) greywacke and interbedded, laminated shale (with uniaxial compressive strength from 25 MPa to 150 MPa).

The Crosscourse pit area was known for poor geotechnical conditions, as the convergence of four faults and the pit wall produces a wedge failure zone. A series of failures of up to 100,000 tonnes occurred along the strike of the Union Fault, which consisted of a few major controlling structure. The largest multiple failure was assessed to be 170,000 tonnes (Fig. 4.1). Failure was initiated along a sub-vertical fault, and the wedge slid along a flatter-lying fault. Continued mining after the failure revealed a low-angle fault, which contributed to the failure Szwedzicki (2004).

Geotechnical indicators of this rock mass failure were as follows:

- the convergence of four faults and the pit wall resulted in a wedge failure zone,
- weathered, low-strength, sub-vertical bedding, and
- high groundwater levels.

Geotechnical precursors of this rock mass failure as follows:

- Nine months earlier – a 100,000-tonnes wedge-type failure occurred immediately adjacent to the final failure. The Union Fault initiated failure along the back of the wedge. High rainfall and mining from the toe of the wall at the time of failure were likely triggers of this failure.
- Two weeks – progressive slide/unravelling failure (60 tonnes) occurring along the controlling fault.
- Nine days – tension cracks developed along the Union Fault contact in weathered material above the failure zone, over a 60 m vertical section. Prism movements above the failure zone recorded 18 mm/day until the slope failed.
- Twelve hours – significant rockfalls and unravelling of material occurred along the fault at the back of the wedge.

Possible triggers to the failure:

- high rainfall and surface water infiltration,
- blasting induced crack propagation, and
- instability caused by rock mass failures in adjacent areas.

Figure 4.1 Multiple slope failures along geotechnical structures (Szwedzicki, 2004).

4.3   COLLAPSE OF A HIGHWALL IN AN OPEN PIT

Telfer Gold Mine located in Western Australia was subject to a slope failure of 176,000 tonnes in 1992. The failure occurred immediately after excavation at the toe of the slope. The highwall was planned with 62° batters and 5 m berms. However, by using a cable bolting system, it was considered possible to steepen up the batter angle to 71.5°. More than two years before, a collapse of a 100 m highwall took place nearby. It was expected that the collapse would propagate. Geotechnical investigations concluded that there was potential for a similar mechanism developing to the south over a strike length of up to 500 m. It was recognized that movement at the toe of the slope would be a precursor for failure.

   Below is a sequence of precursors which preceded the final highwall failure (Thompson and Cierlitza, 1993):

- About 13 months prior to the failure – an increase in the rate of movement (0.08 mm/day) was indicated by two extensometers in the toe of the slope.
- In eight to twelve month period – the movement was recorded in additional two extensometers.
- Eight months – inspection of the highwall revealed that cracks had appeared on the surface. Shortly after that observation, faceplates on the cable bolts directly below the cracks began to deform, indicating the cables were beginning to carry significant load. It should be noted that extensometers recognized that failure was occurring almost six months before any visual signs of deterioration in the slope were identified.
- Over two months following the initial appearance of the cracks, further cracks appeared along the strike for a distance of 300 m. In addition, cracks appeared at regular intervals for a distance of up to 160 m back from the crest of the slope.
- Four months – footwall heave was observed at the toe of the slope.
- About three months – there was a marked increase in the rate of movement shown by prisms. The rate of movement almost doubled from around 0.6 mm/day to 1.1 mm/day.
- Two months – additional cracks were observed. This was followed by a small failure of approximately 20 tonnes.
- Three weeks – a rockfall of approximately 600 tonnes occurred. The rockfall, limited in size, occurred between the prism monitoring points.
- Six days – a next rockfall of 200 tonnes.

Following the failure, monitoring continued. Instrumentation outside the failure area showed a dramatic decrease in the rate of movement. Within a few months, several cracks showed negative movements corresponding to cracks closing behind the crest of the slope.

Chapter 5

Case studies of inundations

Inrush into mines or spillage of liquefied waste material have been recorded throughout mining history. Uncontrolled mass movement of water, mud, wet backfill and tailings flooded mines, often resulting in operations closure. An inrush of water, mud or tailings into a mine represent low-probability but high-impact risk. Most common causes of inrushes are surface water or liquefied material entering mines, breaking into old excavations, and movement of liquefied backfill from stopes or fine material from orepasses.

The following reasons were identified for water mud and tailings inrush into mines (Szwedzicki, 2003b):

- Deficiency in mine design
  - Unsuitable mining methods
  - Ineffective protective barriers
  - Errors in design assumptions

- Deficiency of old mine plans
  - Incomplete old plans
  - Absence of old plans
  - Incorrect interpretation of old plans

- Incorrect geological interpretation
- Absence of protective pillars
- Failure to plug boreholes

Large-scale mining requires construction of not only large mines but also large tailings storage facilities and waste rock dumps encompassing millions of tonnes of fine tailings material or fragmented waste rock. Such material when saturated with water (e.g. during heavy rainfalls), can liquefy or be easily eroded. Lists of geotechnical failures of such structures on mining ground is given in Environmental and Safety Incidents Concerning Tailings Dams at Mines, UNDP, 1996, and also in *Tailings Dam Risk of Dangerous Occurrences, Lessons Learnt From Practical Experiences*, Commission Internationale des Grandes Barrages, Paris, 2001. Disasters caused by the failure of storage facilities create not only safety hazards and financial losses but also cause environmental destruction.

The similarities of the circumstances and events leading to the failure showed that lessons which should have been learned from these events had been forgotten. Seven case studies of inundation of water, tailings, backfill and mud were reviewed and indicators prior to liquefaction are discussed. In these studies, geotechnical modes of failure and precursors to instability of these structures were identified in hindsight and are given below.

5.1 WATER INRUSH INTO A COLLIERY

In 1996, at Gretley Colliery, NSW, Australia, a team of miners was developing a roadway in 50/51 panel operating a continuous miner. Gretley Colliery was known as a wet mine, but 50/51 panel was one of the driest panels in the mine. Suddenly, with tremendous force, water rushed into the heading from a hole in the face made by the continuous miner. That machine, weighing between 35 and 50 tonnes, was swept some 17.5 m (Staunton, 1998). The water came from the long-abandoned old workings of the Young Wallsend Colliery, which ceased operations in 1892. The workings of the old mine were full of water, extending to the surface (about 150 m). The Gretley Colliery was working to a plan that showed the Young Wallsend Colliery more than 100 m away from the point of holing-in.

An indicator of a potential rock mass failure was a large accumulation of water in the neighbouring mine. The following are precursors during the period of two weeks preceding the inrush:

- About two weeks before – a statutory report was made: "Large amount of nuisance water in C 7 cut-through". The depth of water in the heading was 0.3 m and the old workings were about 75 m away. At that time, the continuous miner moved to another working place and the heading remained undisturbed.
- Ten days before – it was noticed that the depth of water had risen to 0.6 m. At that time, two separate statements were made: "Amount of water in the panel appeared to be increasing", and "The watermake (*sic*) in the section appeared to be unusual."
- A day before the inrush – it was reported that "Coal seam is giving out considerable amount of water seepage at face C heading". The old workings were only 7–8 m away.
- Cutting through the partition between old and new workings triggered the inrush.

5.2 TAILINGS INRUSH INTO AN UNDERGROUND MINE

Mufulira Copper Mine, Zambia, commenced underground production in 1931 and in 1944 introduced caving methods. The most productive section of the mine was overlain by a large tailings dam. Progressive collapse of the hangingwall led to the formation of a surface depression below the dam. In total over the period of 1933–1975, more than 20 million tonnes of tailings were deposited, forming a large accumulation of saturated slimes (Sandy *et al.*, 1976).

A chimney cave developed beneath this accumulation and allowed the tailings to flow into the mine through ingress points on the 580-metre level. In 1970, an estimated 680,000 m^3 of liquefied tailings inundated the mine. The sequence of significant geo-technical events prior to the disaster is given in the Final Report, The Commission of Inquiry – Mufulira Mine Disaster (Mufulira Mine, 1971). A chimney cave developed beneath this accumulation and allowed the tailings to flow into the mine through ingress points on the 580-metre level. An aerial photograph taken shortly after tailings inrush is shown in (Fig. 5.1)

With increased production in the 1960s, a geotechnical indicator of the rock mass failure was that the surface depressions increased, and large pools of stagnant water accumulated.

Figure 5.1 Aerial photograph taken shortly after tailings inrush, Mufulira Copper Mine, 1974 (Szwedzicki, 2001a).

Geotechnical precursors to rock mass failure are as follows:

- Twenty-two months before – a sinkhole 60 m in diameter developed 150 m from the later subsidence centre.
- Eight months – unusually high pressure was recorded on the 518 m level which eventually resulted in three sublevel crosscuts having to be abandoned.
- Five months – a mud extrusion was recorded on the 533 m level. The mud was of the consistency of toothpaste.
- Three months – further extrusions of the mud affected five crosscuts.
- Two months – a sinkhole 60 m in diameter developed centrally within the area of the later massive subsidence.
- One month – further mud extrusion.
- Six days – a circular depression some 60 m in diameter appeared on the surface (about 60 m from the centre of the later subsidence).
- Five days – a small sinkhole 9 m in diameter appeared within the depression.

A possible trigger to the failure was as follows:

- Stope blasting during a shift prior to collapse.

In 15 minutes or less, the tailings entered the mine. The Commission of Inquiry concluded that the disaster was caused by faulty operational procedures together with a deficient tailings disposal practice and a drainage scheme, all of which, although sound in isolation, became a dangerous combination.

5.3 BACKFILL LIQUEFACTION AND INRUSH INTO A MINE

A precursory pattern prior an event was reported at the Bronzewing Gold Mine, Western Australia, where a backfill barricade collapsed, and a large amount of fill material poured into the mine. An open stope was filled with uncemented hydraulic fill pumped from the surface to the top of the stope. It was then allowed to fall to the bottom, gradually filling the stope. The fill was held in place by a barricade. The barricade was 5.5 m × 5.5 m and the wall was 400 mm thick.

With an over 50 m height of saturated backfill, amounting to approximately 22,000 m³, the barricade collapsed. When the barricade wall collapsed, backfill material and water flowed through the mine.

The investigation revealed that the inundation was caused by inadequate drainage of the backfill material and the failure of the barricade. A factor that could contribute to inadequate drainage was larger than stipulated percentage of fines passing 10 microns. Inadequate draining resulted in water ponding above the fill.

5.4 MUD INRUSH RESULTING FROM COLLAPSE OF A CROWN PILLAR

The Balmoral Mine, Ontario, used a shrinkage mining method to extract gold ore between the 100 m and 200 m levels. In 1980, the surface crown pillar above 2–7 stope of the Balmoral Mine caved in. In a few hours more than 100,000 tonnes of mud and water engulfed the mine. When the flow of mud stopped, a crater 20 m deep and 80 m in diameter developed on the surface (Koivu, 1982).

The reported precursors are as follows:

- A year before:

  - The cave-in in poor ground conditions led to the abandonment of excavation of exploration drift directly above 2–7 stope. Poor ground conditions resulted in a decision to leave an 8 m crown pillar in 2–7 stope.
  - Water inrush through diamond drill holes, which flooded the mine.

- Five months – water reappearing in the mine, resulting in three days' production loss.
- Three months – a fall of ground and the crown pillar reduced in size to less than 3 m. The grade of ore was decreasing, indicating substantial dilution.
- Eight days – rock noises were heard in the stope.
- Seven days – additional rock noise and a fall of ground, which produced an air blast.
- Five days –because of deteriorating ground conditions resulting in instability of the back and walls, the production from the stope was terminated. The exploration drift above the stope collapsed.
- A few hours – inflow of muddy water from the stope was reported.

5.5   INSTABILITY OF WASTE ROCK TIP

The massive slide of a part of the Merthyr Vale Colliery waste tip engulfed a school and many houses in the South Wales village of Aberfan (Fig. 5.2) in 1966. Sudden loss of cohesion in the waste rock caused rock in the form of a slurry to avalanche for over half of a mile down the valley side to Aberfan. Prior to the disaster, Tip 7 was about 33 m high and contained 250,000 m³ of loosely tipped, uncompacted material which, in its lower parts, had become over the years saturated with water. The material deposited on Tip 7 was tipped across the material which had slipped away from Tip 4 some 22 years before the disaster. Tipping on Tip 7 began in 1958.

Geotechnical precursors prior to the disaster were as follows (Report of the Tribunal, 1967):

- Four to five years – tenants of the fields on the mountainside had to renew fences from time to time as the Tip 7 material crept downwards.
- Three years – there was a major waste rock slip.
- Over two years – a "fairly large" quantity of tip material (about 20–24 m in height) had slipped some 200–300 m, covering a stream and leaving a large bowl-shaped depression on the side of Tip 7. During the two-year period prior to the disaster, the tip advanced about 2.5 m while its toe was advancing ever downwards and across the mountainside.

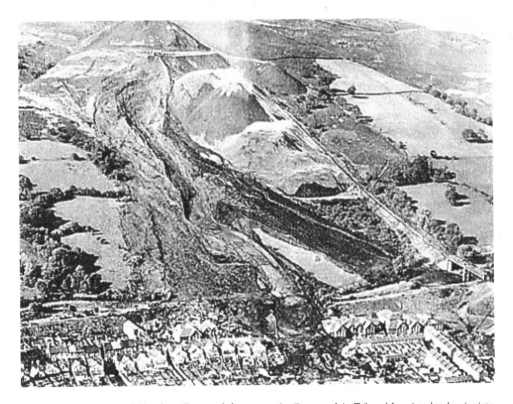

Figure 5.2 Aerial view of Aberfan village and the waste tip. (Report of the Tribunal Appointed to Inquire into the Disaster at Aberfan on October 21st, 1966).

- In the period of one and a half to two years – there were further slipping movements.
- During the last six months – the toe of Tip 7 advanced downwards by some 7 to 10 m.
- During the last three to four months – the toe moved downwards another 7 to 10 m. Credible evidence showed that there was frequent sinking at the top of the cone of the tip material. A few times, during that period, the sinkings were of a magnitude of 3–4 m.
- Two hours – workers arrived at the point of the tip to find that it had sunk by about 3 m and that two pairs of rails, forming a part of the track on which the crane moved, had fallen into the depression.
- About 15 minutes prior, workers found that the point of the tip had sunk another 3 m, so at that time it was beyond question that it was 6 m below its normal level.

Possible trigger:

- Accumulation of water emanating from permeable slopes forming at the base of the tip.

5.6   INSTABILITY OF TAILINGS DAM

In 1994, a 31 m high tailings dam upslope of Marriespruit in South Africa failed. The dam failed a few hours after 30–50 mm of rain fell in approximately 30 minutes. The failure resulted in some 600,000 m^3 of liquefied tailings flowing though the town. The tailings flowed for a distance of about 4 km. Just before the failure, the freeboard was 0.3 m (Wagener et al., 1998). The wave of tailings was about 2.5 m high when it reached the first row of brick-built houses, about 300 m downslope of the dam. Some of the houses were swept off their foundations, while walls were ripped off others.

   Geotechnical precursors to instability were as follows:

- The northern wall of the dam had been showing distress for a number of years in the form of seepage and sloughing near the toe. Attempts to stabilize the wall led to the construction of a drained tailings buttress. Continued sloughing resulted in a decision to discontinue tailings disposal about 15 months prior to the disaster.
- Sloughing at the toe continued.
- Landsat satellites measured reflected solar energy three weeks before the disaster and revealed wet conditions below some parts of the north wall.
- Shortly before the disaster, slips occurred on the lower slope of the dam immediately above the buttress. One person reported seeing water flowing over the top of the slimes dam wall.

Trigger to the failure:

- As a result of heavy rain, water started overtopping at the lowest point of the north wall. This exacerbated small slip failures on the lower slope of the dam and resulted in a series of slip failures progressively moving up the slope leading to massive overall slope failure that released the flow slide.

The most likely chain of events leading to failure was rain flowing into impoundments which already contained some 40,000 m^3 to 50,000 m^3 of water pumped from the mill. That resulted in overtopping where the pool touched the lowest point of the wall. The spilling water eroded the loose tailings, which in turn led to small slip failures, ultimately leading to a massive overall slope failure that released the flow slide. The solidified mud-covered area was over 500,000 m².

5.7   PROGRESSIVE FAILURE OF COAL REFUSE DAM

In 1972, a coal refuse dam failed, flooding the Buffalo Creek valley which resulted in fatalities and caused extended flood damage. The coal refuse, after processing, was stored behind dams forming a so-called pool. The dams were embankments that blocked a watercourse and which therefore impounded water and other materials. The embankments were built by dumping coal-processing refuse which was not compacted. Coal refuse was pumped to dams in the Buffalo Creek valley. Dam no. 1 was at the lowest location in the valley and dam no. 3 the highest. Dam no. 2 was located 200 m upstream from dam no. 1. Dam no. 3 was nearly 200 m wide. The pre-failure volume of the refuse bank was estimated at 1.6 million m³ (US Department of the Interior, 1973).

The failure was initiated by overtopping. The overtopping was defined as the progressive erosion of the embankment beginning with water from the reservoir flowing over the top of the embankment at its lowest point. As the erosive action continued, the initial channel was widened by undercutting the sides, thereby increasing the quantity of flow.

The initial failure occurred downstream of dam no. 3 and consisted of massive slide movement involving approximately 100,000 m³ of material. This slide displaced pool no. 2 sediments. The initial flood wave then continued downstream, overtopping and destroying dam no. 1 and consisted of a massive slide movement involving approximately 350,000 m³ of sludge and embankment material (Fig. 5.3). The length of the initial failure which occurred in the downstream section of dam no. 3 was about 90 m. The mobilised material displaced sediments in the pool no. 2. The total failure took about 15 minutes.

Figure 5.3 Aerial view of Buffalo Creek area after the event, 1972. (US Dept. of the Interior, Washington, D.C. 1973).

During the five years before the disastrous overtopping of dam no. 1, the following occurrences took place:

- Water from dam no. 3 was continuously seeping into pool no. 2.
- Three years before, a large slump caused by foundation failure occurred on the embankment of dam no. 3.
- Two years before, embankment of dam no. 2 "cracked down and slumped". This was repaired by dumping more refuse into the void created by the slump.
- Two years before, another failure occurred on the downstream face of dam no. 3. This failure was reported to be 50–70 m wide and about 10 m thick.
- About one and a half to one year before, "boils" of black water were appearing in pool no. 2. These boils were evidence of foundation piping and constituted a warning signal.
- A year before, an aerial photograph showed four slides on the downstream of dam no. 3 and one on the upstream face.

The indicators to potential failure were as follows:

- Steep slopes. Slopes of the dams were between 25 degrees and 39 degrees but locally they were over-steepened to 45 degrees.
- Lack of material compaction.
- Variability in embankment material properties.

A potential trigger:

- Ten hours before the failure was a rainfall that totalled about 50 mm.

5.8   FAILURE OF TAILINGS DAMS TRIGGERED BY EARTHQUAKES

Analysis of all well-documented cases of geotechnical failure prove that prior to failure, there must be some structural damage and a trigger that causes material mass movement.

In cases of rock mass or mining structures that underwent substantial damage which reached a precarious state of stability but still could transfer some stresses, a small trigger like small blasting or rain could instigate failure. However, it also has been documented that rock mass or structures which were in relatively stable conditions suffered substantial damage from large-scale triggers like earthquakes.

Failures in many mines or surface structures were caused by earthquakes or tectonic plate movement. One example of a tailings dam failure initiated by an earthquake is El Cobre. In 1965, a devastating earthquake of magnitude 7 to 7¼ on Richter scale shook the central zone of Chile with devastating effect. Two of three dams at El Cobre were completely destroyed and more than two million tonnes of tailings flowed into the valley, travelling more than 12 km in a few minutes. The failure of both dams occurred at the same time as or immediately following the earthquake. At the time of the failure, the dams were more than 18 months old and had begun to consolidate. The impact of the earthquake liquefied the material to such an extreme that waves, up to 8 m high, were produced on its surface (Dobry and Alvarez, 1967).

Another example of failure instigated by an earthquake is the collapse of a dam at Minas Gerais (Morgenstern et al., 2016). In 2015, the 110 m high tailings dam in Minas

Gerais collapsed with some 32 million m³ of tailings, representing 61% of the impound-ment contents, avalanching down the valley. The iron ore tailings were transported in a slurry consisting of sand and slimes. Slimes consisted of the clay-sized particles that remain sus-pended and slowly settle in standing water to produce a low-permeability material. The sands were deposited to form a buttress or "stack" that retained the slimes discharged sepa-rately behind it.

It was reported that the tailings dam suffered geotechnical damage prior to collapse:

- In 2009, large seepage flows carrying fines appeared on the downstream slope above the main underdrain, which is symptomatic of the process of piping or internal erosion.
- In 2010, inspections revealed cracking and structural damage from foundation settle-ment and construction defects. A vortex appeared in the reservoir showing where tail-ings and water were entering.
- In 2012, a sinkhole appeared in the tailings.
- During 2013, subsequent surface seepages began to appear on the left abutment at var-ious elevations and times. Cracking began appearing at several locations at the left abutment.
- In 2014, a series of cracks much more extensive than anything that had occurred in the previous years were discovered that extended behind the dam crest, emerged at the toe and encompassed most of the slope.
- In 2015, three seeps occurred. The first such incident occurred in March, followed by another seep in June. A third seep on November 15 was accompanied by slumping of the slope.

On 5 November 2015, two blasts occurred at a nearby mine within seconds of each other. Some 90 minutes later, three small-magnitude earthquakes occurred over a period of four minutes. Pre-failure earthquakes were of magnitude 1.8 to 2.6 and occurred almost directly beneath the iron ore deposit.

About two hours later, the dam started to collapse. A cloud of dust had formed over the left abutment, and cracks were forming. The slope was beginning to undulate "like a wave" as if it were "melting", bringing the dam crest down, looking like an avalanche of mud-like tailings cascading down. Eyewitnesses described slope movement propagating "from the bottom up" on the lower benches, not from the crest down. The tailings that had been solid ground just minutes before transformed into a roiling river, overtopping but not breaching a downstream dam, then entering a town shortly thereafter en route to its ultimate destination in the sea.

The following potential failure modes were considered:

- overtopping,
- internal erosion,
- foundation or embankment sliding, and
- liquefaction.

All but liquefaction was ruled out as being inconsistent with physical evidence and/or eye-witness accounts.

The initiation of a flow required not only the presence of saturated tailings but also a trigger mechanism to initiate the process that mobilizes undrained shearing and hence flow sliding.

Chapter 6

Effect of discontinuities on the initiation of failure process

A geotechnical specification of the rock mass requires information on the mechanical properties of the intact rock. The mechanical properties of a rock mass are affected by discontinuities such as planes of weakness, mineralogical variations, bedding planes, cracks, flaws, joints, etc. On a macroscopic scale, the discontinuities are characterised by distinctive joints, and the tensile strength of the rock mass is often considered to be zero. On a laboratory scale, there are microscopic discontinuities like micro-defects, intergranular cracks and micro-flaws that require special detection methods. These microscopic discontinuities affect the behaviour of intact rock and consequently the rock mass (Szwedzicki and Shamu, 1996).

Even when rocks of identical lithological composition are tested, the existence of microscopic discontinuities results in variation in the value of mechanical properties, especially on the uniaxial compressive strength. Whilst it is widely accepted that discontinuities affect mechanical properties of rock and the existence of micro-cracks and their effect on failure propagation were investigated by numerous authors (e.g. Farmer and Kemeny, 1992; Peng and Johnson, 1972; Horii and Nemat-Nasser, 1985), current mine design practices and standards do not take into account the effects of these discontinuities on the mechanical properties of the rock.

6.1 MODES OF FAILURE OF ROCK SAMPLES

Under high stress, microscopic discontinuities determine the failure mode. The location, orientation, size, density and extent of discontinuities contribute to different modes of failure of the rock. Modes of failure of cylindrical rock samples have been observed and classified (e.g. Reinhart, 1966; Paul and Gangal, 1966; Fairhurst and Cook, 1966). However, these observations are not considered when interpreting the strength of rock or rock mass. Various modes of failure of mine pillars were classified by Brady and Brown (1993).

For hard and brittle cylindrical rock samples, five distinct modes of failure were identified: simple extension, multiple extension, multiple fracturing, multiple shear and simple shear; these modes are depicted in Figure 6.1 (Szwedzicki and Shamu, 1999).

Rock tested under load generates forces of shear and tension. Consequently, rock fails in shear, in tension or in any combination of these.

6.1.1 Simple extension

The simple extension mode (vertical splitting, axial cleavage) denotes a failure along a plane parallel to the direction of compression. The simple extension failure doesn't happen frequently, and such a failure mode may suggest that the sample was relatively free of microscopic discontinuities or that discontinuities were aligned with the principal stress.

Simple extension   Multiple extension   Multiple fracturing   Multiple shear   Single shear

Figure 6.1 Modes of failures of cylindrical rock samples.

6.1.2 Multiple extension

Where two or more fractures run parallel to the long axis of the sample, with fracture perpendicular to that direction, multiple extension failure takes place.

6.1.3 Multiple fracturing

Multiple fracturing involves sample disintegration along many planes at various angles. This type of failure of the specimen is often dynamic and violent, with a large amount of energy being released. When tensile failure is predominant, most of the disintegration planes are in vertical and perpendicular directions to the loading force. When shear forces are predominant, the sample disintegrates along planes inclined and intersecting the mid-height of the sample, for example hourglassing or cone failure.

6.1.4 Multiple shear

When fracturing takes place along two or more planes situated obliquely to the direction of compression, the mode is called multiple shear. The shear surfaces can be identified by the dust left behind when fracturing occurs.

6.1.5 Simple shear

The single shear failure involves one parallel shearing plane (zone) situated at an oblique angle to the direction of maximum compression. The shear planes usually develop across an unconfined part of the sample. Single shear may include shear failure resulting from uneven loading of the sample. This happens when a single shear commences from the top or bottom of the sample and progresses outwards. The maximum testing load for simple shear is often low compared with other failures, since the failure plane is often associated with a discontinuity or weak vein material.

When testing rock samples, in almost every sample, several modes could be noticed. However, it appears that rock samples in the uniaxial compressive test predominantly fail in shear (simple or multiple).

Current laboratory testing practices do not consider the effect of microscopic discontinuities on the mechanical properties of the rock samples. When samples of identical lithological composition are tested, the existence of discontinuities results in variation in the
values of mechanical properties, especially on the uniaxial compressive strength. Large
variations in values of uniaxial compressive strength obtained by conventional means of
determining rock strength are common. The measures used to improve the accuracy of
results include:

- testing a statistically significant number of samples,
- improving the precision of a sample preparation, and
- testing many samples and disregarding values that are anomalously high or low.

Many laboratory tests produce results that do not necessarily reflect the values of mechanical
properties of intact rock material. In current practice of rock engineering design, the results
of testing are arbitrarily scaled down to account for "scale effects" (e.g. Bandis, 1990).

The fact that discontinuities affect mechanical properties of rock is widely acknowledged
(e.g. Hoek and Brown, 1980; Jumikis, 1983), and various types of rock material discontinuities have been classified (e.g. Vutukuri *et al.*, 1974). Although various modes of failure for
cylindrical samples have been observed and classified by numerous investigators, these are
not considered when interpreting the results of strength testing.

The non-destructive testing technique can be used to inspect samples prior to and after
mechanical testing (Gardner and Pincus, 1968). The use of the fluorescent penetrant dye to
detect microscopic rock material discontinuities proved successful and makes it possible
to clearly define discontinuities that would normally have gone undetected. The technique
allows for visualisation of closed cracks, bedding planes and bonded planes of weakness
which have surface traces. Observations under ultraviolet light of rock fragments after
mechanical testing helps to determine whether samples failed along microscopic discontinuities or along new, stress-induced failure planes.

The effect of microscopic discontinuities on the mode of failure and propagation of the
failure planes was assessed by inspection of the rock samples and an analysis of the treated
samples under ultraviolet light before and after destructive testing. From a study of the
discontinuities revealed prior to destructive testing and the rock fragments produced after
failure, three types of fracture propagation were distinguished:

- fracture initiation at imperfection and a subsequent propagation through intact rock
  (Fig. 6.2),
- fracture initiation on many discontinuities and propagation throughout intact rock material, and
- fracture formation and propagation along detected planes of weakness (Fig. 6.3)
  (Szwedzicki and Shamu, 1999).

In cases when the discontinuities were located at the end sections of a sample, failure took
place in a single or a multiple shear mode. Figure 6.2 shows a dolerite sample that failed in
shear mode along planes of weakness situated near the end of the sample.

Where rock material discontinuities were oriented parallel or subparallel to the load
axis, failure tended to occur in a vertical splitting mode, as shown on a quartzite sample
in Figure 6.3.

Figure 6.2 Failure along obliquely oriented planes of weakness – inspected under normal light (top); inspected under ultraviolet light before testing (middle); inspected under ultraviolet light after testing (bottom) (Szwedzicki and Shamu, 1999).

Figure 6.3 Vertical splitting of a sample along a detected plane of weakness inspected under normal light before testing (top); inspected under ultraviolet light before testing (middle); inspected under ultraviolet light after testing (bottom) (Szwedzicki and Shamu, 1999).

Discontinuities oriented perpendicular to the load axis appeared to have less effect on the mode of failure. The other factor that influenced the mode of failure was the relative size and extent of discontinuities. It was observed that if the detected discontinuities were persistent, fractures tended to propagate along them forming planes of failure. Where discontinuities were non-persistent, the failure often initiated at these discontinuities but propagated in a random direction.

6.2   EFFECT OF DISCONTINUITIES ON STRENGTH OF ROCK SAMPLES

The uniaxial compressive strength of a rock sample is a function of the mechanical properties of the intact rock and of the mechanical properties of microscopic discontinuities and hence is related to the mode of failure. To quantify the effect of the discontinuities on the failure mode of samples of different lithologies, a dimensionless parameter was introduced. This parameter was defined as the compressive strength normalised by the mechanical property that is least affected by the microscopic discontinuities. The uniaxial compressive strength was normalised by the tensile strength of the rock determined by the Brazilian method. During the Brazilian tensile strength test, fractures propagate across a diametrical plane parallel to loading, and the probability that material discontinuities are aligned exactly with the plane of failure is small. Under such conditions, the tensile (Brazilian) strength of the sample can be regarded as representative strength of the intact rock, i.e. the strength value was not affected by the existence of microscopic discontinuities. For practical purposes, it is often assumed that the ratio between compressive and tensile strength for rock samples is 10:1 (Stacey and Page, 1986). However, Griffith crack theory predicts that the ratio of uniaxial compressive strength at crack extension to the uniaxial tensile strength will always be 8:1. The study revealed that the ratio depends on the mode of failure and can vary from 1 to more than 30. Low values of the ratio of compressive to tensile strength indicate that the failure of the sample compression was influenced by discontinuities, and the compressive strength predominantly represented the properties of the microscopic discontinuities. High values of the ratio indicate that under compression, the sample fails across the intact rock and that the compressive strength represents the properties of the intact material. Figure 6.4 shows the relation between the mode of failure and the dimensionless strength ratio. Simple and multiple shear failures occurred by shearing or sliding. This type of failure occurred along extensive discontinuities that were oriented at an oblique angle to the loading direction. Where distinct microscopic discontinuities, such as bedding planes, or bonded joints, were detected, it was noted that a shear failure propagated along these features. The compressive strength values for single shear mode tended to be the lowest for each rock type. For such mode, the compressive to tensile (Brazilian) strength ratios ranged from 1:1 to 16:1, with an average of 6:1. It is interesting that some samples tested in compression failed in shear along planes of discontinuities at a very low load despite high values of tensile (Brazilian) strength. In multiple shearing, the compressive to tensile strength ratios were generally higher than those for simple shear failure and ranged from 5:1 to 20:1, with an average of 10:1. The low values of the ratio, less than 10:1, indicated that the compressive strength obtained when samples fail in single or multiple mode may not be representative of the intact rock strength. The strength values in this mode can be used to estimate the strength along microscopic discontinuities within the samples. In hard and brittle samples where discontinuities were randomly

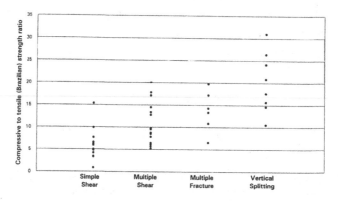

Figure 6.4 Predominant mode of failure vs compressive/tensile (Brazilian) strength ratio (Szwedzicki and Shamu, 1999).

oriented, the predominant mode of failure was multiple fracturing. Under peak load, the samples disintegrated dynamically into small fragments. The compressive to tensile strength ratio ranged from 7:1 to 21:1, with an average of 15:1. The vertical splitting failure mode has been noticed in samples where microscopic discontinuities were not detected or where they were detected but were located parallel or subparallel to the load axis. The developed fracture planes were aligned with the direction of maximum compression (Fairhurst and Cook, 1966). The compressive to tensile strength ratios vertical splitting varied from 10:1 to 31:1, with an average of 20:1, and the strength value could be taken as representative of intact rock (Fig. 6.4). Non-destructive testing of rock samples prior to testing can be used to select samples that are free from microscopic discontinuities. The existence of microscopic discontinuities requires that "scale effects" factor, i.e. strength reduction with increase of specimen size, is used for design purposes. The larger the specimen, the higher the probability that the microscopic discontinuities will affect the strength.

It was found that the discontinuities influence the mode of failure. Shear failure was most common where discontinuities were oriented at an acute angle to the load direction and when discontinuities were revealed at the top or bottom part of the sample. When the discontinuities were located at the mid-height part of the rock cylinder, the multiple fracturing mode was predominant. Vertical splitting took place where the discontinuities were not detected.

The compressive to tensile (Brazilian) strength ratio and the mode of failure gave an indication to the origin of the fracture propagation and must be analysed before the results can be used with a high degree of confidence in geotechnical analysis.

The non-destructive method of sample inspection, when used in conjunction with an analysis of the mode of failure and the rock strength, can assist in explaining the wide scatter of results obtained from the same sample population.

Chapter 7

Behaviour of rock mass prior to failure

Prediction is very difficult, especially about the future.

Niels Bohr

Throughout mining history, miners have recognized certain features of rock mass behaviour as tell-tale signs of impending failure and collapse. In surface mining, "reading the ground" includes observations of crack formation, whereas in underground mining, convergence of excavations or deformation of ground support is usually monitored. Various manifestations of rock mass response to mining can be identified in all phases of mining activities.

Open pits, underground mines, tailings dams and waste rock dumps are mining structures which can suffer as result of geotechnical damage. These structures under stress (which can be mining-induced or brought about by external conditions) may change their response. As a result of the stress, the structures can suffer damage although they can maintain integrity and perform their function.

7.1 PRE-FAILURE WARNING SIGNS

Analysis of documented case studies of mine scale rock mass indicates that geotechnical failure of mining structures doesn't happen without warning (Szwedzicki, 1999a). Structural damage and progressive failure is manifested by the presence of geotechnical warning signs (indicators and precursors) and can be instigated by triggers (Szwedzicki, 2001a). Indicators and precursors leading to local damage and consequently to mine scale failure are shown in Figure 7.1.

7.1.1 Indicators

An indicator is defined as a sign, a state or a contributing factor that points out or suggests that the rock mass may be prone to damage or failure. Usually indicators suggest that the properties of the effected rocks are different from the surrounding rock mass. In general, potential failure is indicated by geotechnical features or mining operational factors.

Geological indicators may include:

- geological disturbance in the form of folds and dykes,
- layers of weak soil or rocks,
- gas-absorbing formations,
- ground discoloration, e.g. resulting from weathering, and
- moisture.

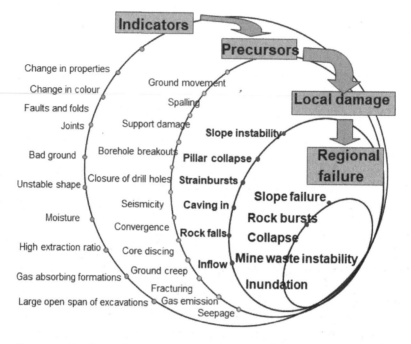

Figure 7.1 Sequence of rock mass behaviour leading to mine scale failure (Szwedzicki and Shamu, 1999).

Geotechnical indicators may include:

- structural features such as faults, shear zones or slickensided planes (Fig. 7.2),
- change in the mechanical properties of the rock mass,
- extrusion of joint fillings,
- water seepage,
- "dog earring", and
- poor ground conditions, e.g. jointed blocky ground (Figs. 7.3 and 7.4).

Figures 7.3 and 7.4 show the same location in a crosscut. Figure 7.3 shows the back of the crosscut with a visible formation of cracks in the shotcrete. These indicators were not recognized as forming a wedge. A few days later, moisture appeared on the shotcrete and the wedge detached from the back resulting in fall of ground (Fig. 7.4).

Ground reinforcement indicators are as follows:

- corrosion or deterioration of support and reinforcement,
- shotcrete cracking,
- breaking of weld mesh,
- damage to plates on bolts, and
- concrete walls fracturing.

Figure 7.2 Slickensided planes facilitating failure of a slope (Szwedzicki and Shamu, 1999).

In addition to geological indicators, analysis of operational indicators can draw attention to possible rock mass structural damage. Such operational indicators include:

- large open spans of underground excavations,
- unstable shapes of underground excavations,
- old excavations near mining activities,
- a high extraction ratio,
- blast damage,
- an accumulation of water in nearby excavations,
- a large body of water or tailings above underground excavations,
- steep slopes in open pits, and
- undercut slopes.

7.1.2 Precursors

A geotechnical precursor (a tell-tale sign) is a state or behaviour that suggests that the geo-technical structure of the rock mass has been damaged prior to possible failure. Precursors, including results from geotechnical instrumentation, warn of the development of excess ground deformations or high stresses. No single precursor may denote structural damage or failure, but many are reported during the process of damage and failure. Observation and monitoring of precursors can give an indication of the scale of structural damage, i.e. local (excavation or level) or mine scale.

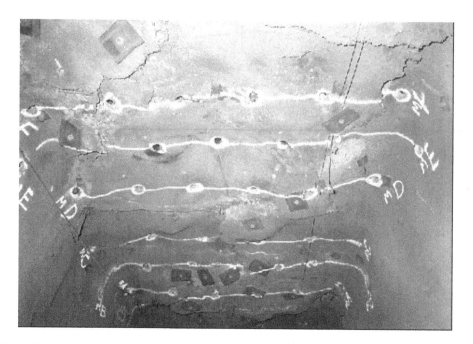

Figure 7.3 Visible cracks formed at the back of a crosscut.

Figure 7.4 Wedge failure along cracks (visible prior to fall on Fig. 7.3.).

Local stress concentration may result in local structural damage. The behaviour of the rock mass on such a scale can be observed at the surface of the excavations. Local scale rock mass damage is predominantly manifested by the following precursory behaviour:

- cracking rock around mine openings,
- unravelling of rock mass (Fig. 7.5),
- spalling of rock from the walls of excavations (Fig. 7.6),
- extrusion of joint fillings,
- roof sagging (Fig. 7.7),
- local rockfalls,
- hang-up rock on mesh (bulging) (Fig. 7.8),
- slabbing,
- joint dilation,
- appearing of moisture and water seepage (Fig. 7.9),
- fracturing of walls of excavations,
- pillar yielding,
- overbreak in excavation corners,
- strain bursts which miners call "popping" or "spitting" as small fragments fly from the rock face, and
- hollow "drumming" behind shotcrete support.

Figure 7.5 Unravelling of rocks from a crest of an open pit (Szwedzicki and Shamu, 1999).

Figure 7.6 Spalling from the hangingwall of an excavation (Szwedzicki and Shamu, 1999).

Figure 7.7 Sagging and shearing of a layer in the roof of an excavation.

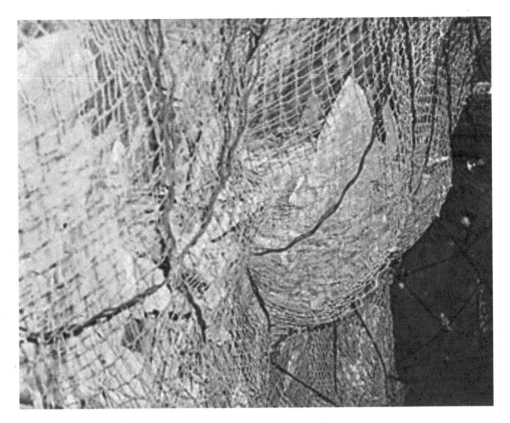

Figure 7.8 Bulging of detached rock fragments behind the mesh.

Figure 7.9 Appearance of moisture through damaged shotcrete.

Certain precursory behaviour indicates that the damage to the rock mass has taken place on a level scale. Such damage is predominantly manifested by:

- pillar punching of the hangingwall or footwall (which may result in floor heaving or roof guttering,
- a need for continuous barring down in areas of loose rock,
- the failure of many pillars,
- core discing,
- borehole breakouts (Fig. 7.10),
- the deformation and closure of drill holes,
- movement across shear zones,
- rock noises,
- hour-glassing of pillars (Fig. 7.11),
- crack propagation (often between levels),
- buckling of layers,
- ground squeeze or bulging,
- creep,
- support damage
- the emission of gases such as carbon dioxide or methane, and
- change in excavation shape (Fig. 7.12).

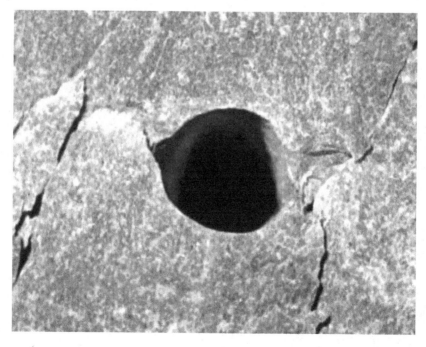

Figure 7.10 Signs of high stress – a borehole breakout and extension cracks (Szwedzicki and Shamu, 1999).

Figure 7.11 Sign of high stress – disintegration of high-strength quartzite in the side of an excavation (Szwedzicki and Shamu, 1999).

Figure 7.12 Change in shape of a circular winze.

Certain precursory behaviour indicates that damage has progressed for a larger distance. Mine scale damage can be manifested by:

- cracks near the crest of the slope (Fig. 7.13),
- movement across faults or sets of joints,
- seismic activity, including rockbursts,
- hangingwall caving,
- surface water disappearance,
- deterioration of ground conditions in part of a mine,
- ground movement (Fig. 7.14)
- formation of depressions on the surface that commonly result in water ponds,
- surface subsidence,
- overdraw from stopes,
- fill migration,
- an increase in water inflow or a change in the water table,
- floor heave at the toe of the slope in open pits,
- wall slumping at the bottom of an open pit (Fig. 7.15), and
- bulging (outward and upward) near the toe of slopes.

Figure 7.13 Progressive opening of cracks at the periphery of an open pit (Szwedzicki and Shamu, 1999).

Figure 7.14 Ground movement at the periphery of an open pit.

Figure 7.15 Slumping of weak rock at the bottom of an open pit.

It is interesting to note that "old miners" used to classify rock noises for assessing the stability of rock mass:

- whispering ground – single cracks of low intensity (indicating local excavation scale damage),
- talking ground – regular cracks of increased intensity (indicating local level scale damage), and
- shouting ground – "big bangs" or "explosions" which were interpreted as "run for your life" situations (indicating mining scale damage and possible collapse).

7.1.3 Triggers

A motion of low-probability of high-impact events stipulates that we don't plan for low-probability events because any one of them is unlikely to occur. And yet the odds of at least one of these events occurring is high – we cannot predict triggers, but we should expect them and foresee the consequences. A geotechnical trigger is any stimulus that impacts rock mass behaviour. Usually triggers appear suddenly and unexpectedly. A trigger can be internal or external. External triggers come from the environment, like water inflow due to heavy rainfall or earthquakes. Internal triggers come from our mining activities, for example, blasting, undercutting slopes or reducing the size of underground pillars. Reduction in mechanical properties of the rock mass below a critical level, for example, due to weathering or water, can be treated as an internal trigger. External triggers usually are unexpected, but internal triggers could be anticipated. We can argue that external triggers are beyond our control while internal triggers can be controlled by managing our mining activities. The failure is usually initiated or exacerbated by triggers like rainfall, blasting or seismic activities.

Most rock mass collapses, especially surface crown pillars and open pit slopes, happen after rain or result from water accumulation. A common cause of collapse is infiltration of water due to heavy rain, rain in areas with large accumulations of snow or rain after prolonged periods of drought. Water leads to loss of strength or washing out of critical binding or key material. Water can wash away joint fills, thereby allowing severe water inflow, free block movement and the transport of fine soil particles from the overburden. Groundwater may cause high pore or high hydraulic pressure in joints that may adversely affect stability. Groundwater may cause rapid deterioration of geotechnical properties due to susceptibility to moisture deterioration, i.e. it can soften the rock mass, reduce its bearing capacity and precipitate its failure.

Many mining structure collapses are triggered by mining activities. The most common is structural damage resulting from blasting. Other triggers include overbreaking due to deficient blast design or poor quality of excavation techniques, for example, overbreaking or undercutting. In extreme situations, the movement of heavy equipment can trigger uncontrolled ground movement (e.g. failure of surface crown pillars).

In the mine literature, it can be found that collapses of mines were also triggered by natural seismic events. In mines that are sited in tectonically active areas, it is also possible that collapses could be triggered by earth crust stress changes due to tectonic plates movement.

Severely damaged or partially failed rock mass can still be in a stable condition. It is interesting to note that a certain delay is observed between triggers and subsequent failure. It is often noticed that ground falls in mines take place a few hours after blasting. In a few cases, massive collapses took place a few days after mass blasting. It has also been reported that the collapse of surface crown pillar or substantial slope failures in open pits usually take place a few days after heavy rainfalls.

7.2   SEQUENCE OF PRECURSORS TO ROCK MASS FAILURE

Geotechnical precursors identified through observations and geotechnical monitoring indicate structural damage to the rock mass. Analysis of documented case studies of large-scale rock mass collapse indicates that it may take years before failure occurs after initial structural damage. In the case of underground mines, the size of the collapsed material was many times the dimension of the mining excavation; the collapse progressed to the surface causing subsidence of large surface areas.

Despite fragmented reports, which in most cases were prepared by personnel with little geotechnical experience, a certain trend can be identified. There is a progression in time, intensity and location of recorded precursors. In almost all rock mass collapses, some precursors were noticed months or even years before the final occurrence. Precursors can be described as ground response observed at a distance from the centre of collapse. In the collapses analysed, the widely used term "ground response or movement" often was not defined or specified. In general, it can be described as surface cracking, movement along planes of weakness and displacement.

Analysis of case studies of geotechnical disasters of mining structures, which included collapses of rock mass of more than 100,000 tonnes, indicates that the precursory behaviour to rock mass failure has a pattern (Szwedzicki, 2004). The sequence of geotechnical precursors commences with ground deformation leading to deterioration in ground conditions and concludes with uncontrolled ground movement. Once the process of failure is initiated at a local scale, it propagates through to the mine scale. It should be noted that certain precursors can be coupled, for example, core discing can take place at the same time as spalling and convergence, while certain precursors can take place sequentially, for example, methane emission can take place before floor heave and that takes place before rock noises.

In cases of regional rock mass failure resulting in massive collapses, the following succession is observed:

*   Initial precursory behaviour is noticed at the periphery of the site of oncoming failure and that the behaviour can be classified as long-term. Ground movement at the periphery of the site of oncoming collapse is observed years in advance.
*   Over time, the precursory behaviour becomes more localized and can be classified as medium-term. Months before collapse, precursors can be broadly described as a deterioration in ground conditions. Noticeable precursors include surface subsidence, small fall of ground occurrences, floor heaving, roof lowering or damage to mining excavations.
*   Within weeks before collapse, precursors are noticed in close vicinity to the centre of the impending collapse site, the main ones being fall of ground, spalling, individual pillar collapse or pillar yielding. In collieries, emission of gases has been noticed. A change in water inflow can also be regarded as a geotechnical precursor to imminent failure.
*   Immediately before the failure, precursory behaviour is short-term. Hours before collapse, the most common observations are rock noises and large falls of ground. It appears that these precursors take place at the centre of impending collapse.

Based on the observed succession, it is possible to distinguish four phases of geotechnical warning:

- An awareness phase, during which initial precursory behaviour can be noticed. This phase is observed years to months before the collapse. Recognition of precursory behaviour allows for a change in mining operations, and remedial action may prevent impending failure.
- An alert phase, during which precursory behaviour is localized. This phase is observed months to weeks before the collapse. During that phase, orderly actions to minimize losses should be instigated.
- Alarm or evacuation phase, during which precursory behaviour is observed at the brunt of impending collapse. This phase is observed days to hours before the collapse. Recognition of that phase allows for urgent action to prevent heavy losses of personnel and equipment.
- Scram (leaving in a hurry) when there is very short warning (minutes) for the evacuation of workers.

7.3   GEOTECHNICAL RISK MANAGEMENT

Rock mass response is inherently complex and not simply reducible to a simple model. The relationship between precursory behaviour and failure is fraught with uncertainty. Matters contributing to a high level of uncertainty include:

- variation in the properties of jointed rock mass,
- assessments of rock mass properties and behaviour,
- stress/strain-failure relationships for rock mass,
- information on *in situ* stress distribution, and
- an understanding of the effects of blasting damage, water, etc.

Lack of knowledge of these factors can contribute to gross simplification, and the results derived from assessment of rock mass stability must be regarded as indicative only. Forecasting is what we must do for risk management. It comprises interconnected stages: anticipation, avoidance and adjustment.

In terms of a risk management control system, indicators can be considered as hazards, precursors as incidents, and local damage and regional failure as losses.

A geotechnical risk management approach includes identifying the potential for uncontrolled ground movement and delineating the zone of potential rock mass instability. In this approach, it is implied that there is a relationship between precursory rock mass behaviour (including results from geotechnical monitoring) and rock mass failure, i.e. precursors have a statistically significant correlation with subsequent failure. Many geotechnical failures are preceded by precursors that indicate the failure, but it appears that at present, there is no mathematically reliable way to predict when the failure will occur. Furthermore, some local-scale failures may not be preceded by noticeable precursors and, conversely, some precursors may not be followed by failure.

The most common way of assessing the potential for rock mass failure is to use a heuristic approach. This approach of retrospective analysis examines past case studies of geotechnical and mining environments in which geotechnical failures took place and uses them to predict rock mass behaviour in similar circumstances. The geotechnical prediction is generally described as foretelling rock mass behaviour based on observations or monitoring. The analysis of rock mass performance based on past behaviour allows for remedial measures to be taken. Prediction or foretelling consists of the following elements, type of occurrence, location, timing and severity. Based on analysis of precursory rock mass response to mining, it is possible to foretell a type of a geotechnical occurrence, for example, rockfall, inundation. It is also possible to foretell the location, for example, the deepest level in an underground mine or the surface above a large stope. However, it is very difficult, if not impossible, to predict the timing. Unfortunately, based on present geotechnical knowledge, it is impossible to foretell the severity of the occurrences.

In the absence of statistical (or any other) evidence that precursors determine rock mass failure, the precautionary principle can be applied. The precautionary principle states that where there is a possibility of an undesirable event, protective action should be taken in advance of "scientific" proof of the potential event (Harding, 2000). The precautionary principle calls for preventative action even where there is not enough proof that failure will occur. In other words, when geotechnical uncertainty does exist, application of the precautionary principle can prevent unwanted events.

Although we may understand relationships, for example, those between stress levels and the failure of a rock sample in a laboratory, there is no proven relationship between precursory behaviour and final failure, such as between rock noises and failure, or between a local rockfall and subsequent large-scale rock mass instability.

When precursory behaviour to rock mass failure is detected, the original collected data must be carefully reviewed for the following (Call, 1992):

- features missed or not considered (e.g. unknown fault, water seepage),
- misinterpretation of a previously obtained interpretation (e.g. strength/stress analysis),
- verification of results from laboratory testing (e.g. was testing done on dry or wet samples),
- appropriate geotechnical data collection, and
- assumptions of geotechnical conditions (e.g. stresses).

Observation of the geotechnical precursors and application of precautionary principles allows for prevention of rock mass failure associated with mining activity or mitigation of the effect of rock mass failure. It can be achieved by changing mining operations, reinforcing the rock mass or applying special techniques to modify the properties of the rock mass (e.g. grouting or draining).

7.4 MONITORING OF PRECURSORY BEHAVIOUR

The role of geotechnical instrumentation is to aid the decision-making process by identifying possible geotechnical risks and modes of rock mass failure and then taking appropriate steps to control the impact. Geotechnical instrumentation requires an understanding of the generalized time sequence of events leading to failure and recognition of the value of early preventive interventions to reduce undesirable events.

In the failure propagation phase, monitoring can provide information on how rock mass behaviour is changing, so that timely warning or implementation of remedial measures becomes possible.

In the early stage of failure, which is usually caused by structural damage to the rock mass, the most valuable gauge is the use of geotechnical instrumentation to measure displacement, for example, extensometers, crackmeters or surveying methods. Use of such instrumentation can quantify elastic and inelastic ground movement, displacement caused by a fracture opening or strain caused by changes in stress. Displacement should be monitored at the perimeter of the site of anticipated collapse.

Displacement monitoring, although providing valuable information on structural changes in the rock mass, appears to have limited value in providing an indication of the onset of the collapse. There are numerous examples of rock mass which was structurally damaged and underwent substantial displacement but did not collapse, including open pit walls which have been in an unstable condition (creeping) or fractured pillars (in underground mines) which have been yielding for many years.

For structurally damaged rock mass, the best source of information on changes in stability is the monitoring of stress changes. Stress changes should be measured in the centre of the anticipated failure area and give an indication of the behaviour of the rock mass. Lack of substantial stress changes, despite progressive displacement, may be an indication of the creep behaviour of the rock mass. To predict failure, stress change must be related to the strength of the rock mass, and that presents a geotechnical challenge. Although changes in stress may indicate impending failure, they still do not provide information on the timing and the scale of the collapse.

Geotechnical instrumentation allows the analysis of rock mass response by measuring changes of critical values over time. Generally, three types of behaviour can be identified: regressive, progressive and transgressive. Regressive behaviour is observed when, after an initial increase in value, the measured parameters stabilize and don't change. Such a trend indicates that the probability of failure is remote or unexpected. Progressive behaviour is reflected by slow and often "linear" increases in value over time. A progressive trend indicates that structural damage is taking place and the probability of failure can be assessed as probable or likely. However, it should be stressed that slopes of open pits may exhibit progressive behaviour for many years and that this does not necessarily end in collapse (Sullivan, 1993). Transgressive behaviour refers to a rapid growth in measured values, which indicates that structural damage is accelerating. The probability of failure can be assessed as imminent. Figure 7.16 shows a generalised relationship between rock mass behaviour over time and the probability of failure.

Rock mass movement or failure can happen when a sufficient trigger is provided independently to the rock mass behaviour. However, with transgressive behaviour, the increase in values (often exponential) indicates that there is a high potential for failure. It has to be stressed that failure will not happen without a trigger.

To illustrate the importance of triggers in instigating failure, a generalized case is shown in Figure 7.17. Rock mass response to mining was monitored and the value of the measured parameter was increasing in time. At a certain value of the parameter, only a large trigger like an earthquake could instigate failure. In absence of that trigger, the measured parameter would increase in value and the failure could be instigated by energy from mass blasting. If the blast energy wasn't high enough, the rock mass would remain in a stable condition and the values of the parameter would increase even further. Now the failure could be instigated

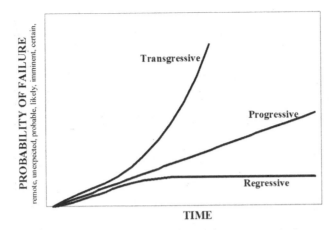

Figure 7.16 Change in values of monitored parameters in time versus probability of failure (Szwedzicki and Shamu, 1999).

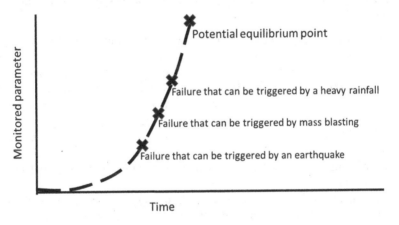

Figure 7.17 Different triggers needed to instigate failure of the rock mass during different stages of its damage.

by heavy rainfall. Water could change the mechanical parameters of the soil or fractured rock mass, leading to failures. If such rainfall is not taking place, the measured parameters keep increasing until another smaller trigger comes, or an equilibrium state is attained. In such a state, even the smallest amount of energy could instigate the failure. Experienced skiers, in high mountain areas, know that a snow avalanche could be triggered by a sound wave only.

The results of the analysis of case studies of geotechnical precursors to ground collapse prove that, in underground and open pit mining, a full range of geotechnical monitoring is required to understand the behaviour of potentially unstable rock mass and to infer rock mass failure. This may include displacement monitoring, water monitoring, stress change monitoring and acoustic emission monitoring. An example of generalized time manifestation of various rock mass, behaviour patterns monitored prior to failure,

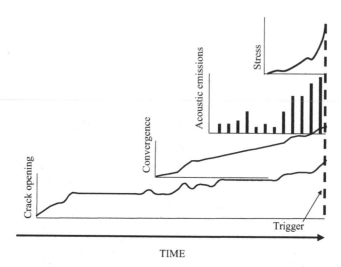

Figure 7.18 Various manifestations of rock mass behaviour monitored prior to failure (Szwedzicki and Shamu, 1999).

is given in Figure 7.18. Expecting ground movement, a crackmeter was installed over an opening fracture. With time, it was decided to install a wire extensometer. Increase in values of measured parameters resulted in the installation of acoustic emission monitoring and subsequently a stress cell to measure stress changes. The readouts were superimposed on a common time scale. All measured parameters were precursory to failure, but it was a trigger which instigated the failure.

It is noticed that rock mass collapses, which take place in hard rock mines, are usually preceded by acoustic emissions. These emissions take place a relatively short time prior to collapse. Monitoring of acoustic emissions may provide a final warning of rock mass failure and imminent collapse.

The development of monitoring and prevention strategies requires planning for the occurrence of the identified hazard. Preparation entails the development of response plans, ensuring that staff know their roles and responsibilities during the event and recovery plans to minimize the operational impact.

Chapter 8

Rock mass behaviour during failure

The definition of rock mass failure does not include the time of failure, that is, the duration of the process of fracturing. Failure can occur over a few seconds through to many years. For dynamic failure like rockbursts, failure duration lasts a few seconds, and it is possible to define the failure point on a stress strain curve indicating loss of peak strength. For progressive or gradual failure that takes a long time without a distinctive point of losing strength, it is often not possible to define a point failure. Progressive failure is covered in Section 8.2.

8.1 CASE STUDIES ON ONSET OF FAILURE

Long-term observations and records indicate that the precursory pattern changes immediately before failure of the rock mass. From these records, "words of experience", or maybe just a deeply rooted belief, it appears that, immediately before failure, there is a short period during which the rock mass shows a reverse behavioural trend. This trend can be referred to as "the calm before the storm" during which the rock mass "gathers momentum" for an impending failure.

"Old colliers" had noted the change in behavioural pattern before gas outbursts and rockbursts. Before gas outbursts, the emission of gases increases, culminating in the outburst. The Royal Commission of Inquiry (Sheehy, 1956) stated that the older generation of miners identified "signs" of impending outbursts – "Of these signs we may quote the following examples . . . sudden reduction of the amount of gas in the working".

Major rockbursts are usually preceded by a smaller seismic event. However, in collieries that are rockburst prone, "old miners" are quite comfortable when the rock mass is "talking", i.e. when there is some acoustic emission. They start to worry when the rock noises dissipate. They believe that energy starts to accumulate before a major rockburst.

In the case of inundation of the Gretley Colliery in New South Wales, it was reported that when approaching the body of water an increase in water inflow was noticeable for a period of two weeks. However, in the formal report into the Gretley disaster (Staunton, 1998), one witness stated that just hours before the inrush, "I had an inspection at the face to see if water was still there and the water had appeared to dry up".

Similar change in precursory pattern was noticed at the Bronzewing Gold Mine, Western Australia, where a backfill barricade collapsed, and a large amount of fill material poured into the mine. Hydraulic pressure behind the barricade was monitored at two-hour intervals. The record of investigation states,

The pressure was rising each time the fill was placed . . . the graph showed a gradual increase until five days before the collapse when pressure stabilized and remained with relatively little variations. About 10 hours before the collapse the readings indicated a sudden and dramatic drop (12.5 per cent). The reason for the fall in pressure was not known.

Again, in the case of crown pillar collapse at the Warrego Mine, Northern Territory (Szwedzicki, 1999b), after a mass blast and before ground collapse 10 days later, "the shaft sump lost water and regained it on a few occasions".

During a failure at Fundao Dam (Morgenstern *et al.*, 2016), a few eyewitnesses described slope movement having propagated "from the bottom up" on the lower benches, not from the crest down. At first, the lower slope progressed slowly forward "like a snake", and then bulged, coming down like a wave.

An interesting phenomenon was recorded during the hearing of the Commission of Inquiry into the Aberfan disaster (Report of Tribunal, 1967). The moment of liquefaction of the waste tip, which was about 33 m high and contained 250,000 m^3 of uncompacted rock material, was seen by a crane driver who stood at the top of the collapsing waste tip. He described it as follows:

I was standing on the edge of the depression. I was looking down into it and what I saw I couldn't believe my eyes. It was starting to come back up. It started to rise slowly at first. I still did not believe it, I thought I was seeing things. Then it rose up after fast, at a tremendous speed. Then it sort of came up out of the depression and turned itself into a wave – that is the only way I can describe it – down towards the mountain . . . towards Aberfan village.

(Report of the Tribunal Appointed to Inquire into the Disaster
at Aberfan, Report of the Tribunal Appointed to Inquire into
the Disaster at Aberfan on October 21st, 1966, page 30)

The above-mentioned phenomena were not tested, measured or even quantified. Their occurrences were often not confirmed, but they represented the observations and/or beliefs of miners. Such phenomena were documented in the reports by commissions of inquiry, without any attempt to explain them.

8.2   CASE STUDIES ON DURATION OF FAILURE

It is very rare that rock mass behaviour during the point of failure is determined. In investigations of large mining disasters, witnesses who were in the vicinity of the seismic events or collapses describe in their own words what they noticed. A few such descriptions are quoted below:

- The Solvey trona mine: "Miners working near the panels reported that they heard a rumbling, a big boom, and then a deafening sound lasting five to six seconds." However, ground falls and rumbling continued for several hours thereafter (Goodspeed, *et al.,* 1995). The fall of ground was accompanied by ammonia and methane emission. The damage included floor heaving, rib slabbing and roof falls (Zipf and Swanson, 1999).

- The Northparkes: A massive caving event took place. The majority of the overlying 200 m of orebody to the surface collapsed. It wasn't a single collapse. During a four-minute period, approximately 14.5 million tonnes of ore fell in the stope (Hebbelwhite, 2003).
- Otjihase mine: The collapse started near the centre of the extraction block, in a form of induced falls of rock. It was a progressive collapse, which resulted in loss of a decline eleven days later. The collapse continued for over two years, by which time an area of approximately 175,000 m^2 was affected. The mode of failure was roof spalling on the up-dip side of the pillars and floor heave on the down-dip side (Klokow, 1992).
- Nowa Ruda Colliery: Just before rock and gas outburst, a loud bang was heard and immediately after about 150 tonnes of pulverised coal was ejected and about 32,000 m^3 carbon dioxide was released (Piatek, 1980).
- Coalbrook Colliery: Three tremors were recorded by four seismological observatories. Initial tremors were single shocks with a duration of a second or two. However, the major collapse resulted in shock lasting for five minutes, showed a continuous record with three maxima (Bryan *et al.*, 1964). Slight vibrations occurred for a few days after the collapse.
- In the case of tailings and waste rock damps liquefaction, the initial process of movement takes a few seconds, but the flow of material could take a few hours.

In all cases of ground collapse, it can be generalised that the point of dynamic failure could be as short as a few seconds. It starts with seismic activities in the form of rock noises, loud acoustic emissions followed by seismic shock. These activities start just before the collapse, reach the highest level during the brunt and slowly diminish over a number of hours or even days. In the case of coal and trona mining, a large amount of gas is emitted during the failure process.

8.3   PROGRESSIVE DAMAGE TO EXCAVATIONS UNDER HIGH MINING-INDUCED STRESS

During a process of progressive non-violent damage, stress transfer through the effected rock mass diminishes slowly. The rock mass can suffer substantial damage and still can be able to transfer some stress, for example, yielding pillars, progressive closure of secondary excavation or movement of open pit slopes. Such damage can take years or months before excavations have to be rehabilitated or abandoned. The excavations progressively lose their functionality, but the rock mass, although damaged, still can transfer stress.

In certain mining situations, the stress can be transferred through an abutment (rock mass in front of the mined-out void), remnants, pillars, or consolidated fragmented rock material in mining stopes. These types of mining-induced stresses are called abutment stresses. The shorter the distance between the extraction front and a mining excavation, the higher the superposition of the abutment stresses and stresses around the excavations. In deep large-scale mines, mining-induced stress can reach very high levels, with the damage accelerating as extraction approaches excavations. High mining-induced stress should be considered not as a value of the maximum principal stress but as a difference between maximum principal and minimum principal stress. Increased mining-induced stress results in stresses which could exceed the strength of the rock mass, leading to progressive and often accelerated damage.

The time from development of an excavation to abandonment or closure is called the lifetime of the excavation. However, more crucial is the stand-up time (time of exposure) counted from the moment of the first sign of high stress (stress transfer) to the time the excavation deteriorated to such extent that it must be re-supported or abandoned.

Damage which can progress in time is a function of mining-induced stresses, i.e. a function of the rock mass properties and the distance from the extraction line to the excavation. Exposure to high stress (relative to the rock mass strength) can cause severe stability problems, resulting in delays to the scheduled production. It has been found that excavations under high abutment stress deteriorate transgressively, i.e. deterioration begins slowly, but then accelerates towards the eventual abandonment. When the production face approaches an excavation, the first signs of high stress can be noticed. From that moment, deterioration in ground conditions can progress quickly and, with an increase in abutment stress, the damage can propagate exponentially, leaving weeks, if not days, before rendering the excavation unusable. To mitigate the effect of abutment stresses and prevent rock mass from damage around excavations, the minimum standoff distance between the advancing production face and excavations must be determined.

The rock mass response to high stress is observed as convergence (closure) or by spalling (enlargement). When stress-dependent closure starts to restrict access (e.g. make it impractical to use mining equipment), the excavations must be rehabilitated. In such cases, the most common support elements are cable bolts with shotcrete and/or steel arches. If not rehabilitated, the excavation must be abandoned. An extreme total closure of a 5.5 m wide drive is shown in Figure 8.1. In the centre of the picture, a few strands of wire mesh, originally used as a component of ground support are visible. It is also common that due to spalling and fracturing, excavations undergo enlargement, as seen in Figure 8.2. The fractures are indicative of stress direction, as the maximum principal stress component is parallel to the deteriorating sides. At down dip below the extraction level, local overstressing is manifested by dog earing, slabbing and the dislocation of blast holes, extensive spalling from the sidewall and backs of the development openings, closure and, in extreme cases, losses of access. Enlargement of the excavation with an increase in mining-induced stress causes progression in the size of the damaged zone, which could result in excavation damage to the extent that it must be re-supported or abandoned for mining purpose. In mining practices, when enlargements are more than 1 m, the excavations need re-support. In such cases, the most common additional ground supports are concrete walls.

Three case studies on the effect of exposure to abutment stress on excavation deformation are discussed below: one in a deep copper mine using open stoping with pillar recovery where stress transfer took place through remnant pillars, the second in a nickel mine using sublevel caving methods where stress transfer took place by down-dip abutment, and the third in a copper mine using block caving method where stress transfer took place through compacted caved-in material.

Figure 8.1 A total closure of a drive originally being 5.5 m wide.

Figure 8.2 A drawpoint enlarged from about 4 m wide to over 7 m wide.

8.3.1   Case study of stress transfer through remnant pillar

In a copper mine, total mining extraction was carried out at a depth of about 800 m. The stopes were 50 m high and 50 m wide. Total extraction was carried out by blast hole open stoping with pillar wrecking and no fill (Broom and Sandy, 1988) with the aim of ultimately caving the rock mass. However, caving was not easily initiated, leading to down-dip transfer of stress.

As mining production advanced, the abutment stress was transferred by remnant pillars. It was noticed that there was a pattern of damage in crosscuts, drawpoints and drives. Initial damage was usually recorded as the undercut of the stope was blasted. Acceleration of damage was noticed as the pillar between stopes was isolated by development of the slot; further acceleration was observed during stope blasting; and deceleration of damage progression as the stope face moved over pillars and excavations were covered by stress shadow.

To assess the effect of stress and time of exposure, a classification of excavation was prepared. The classification was for different stages of development damage for crosscuts, drives and drawpoints. The classes of damage were based on excavation enlargement due to slabbing, spalling or fracturing (Kurzeja, 1992). The following classes were identified:

- Class 1 – no damage due to effect of stress transfer.
- Class 2 – minor damage – enlargement of less than 0.5 m.
- Class 3 – substantial damage – enlargement of less than 1 m.
- Class 4 – extensive damage – enlargement to 1.5 m.
- Class 5 – severe damage – enlargement of over 1.5 m.

Figure 8.3 Damage to a crosscut with the caving line at a distance of about 20 m with exposure time of 30 months (Szwedzicki, 1989a).

For minor damage, some minor support work, like spot bolting or shotcrete, was recommended. For substantial damage, full re-support was applied. For extensive damage, rehabilitation was required, or excavation was declared as restricted access and barricaded. For severe damage, including total collapse, passive support such as timber packs or concrete walls were required, or the excavations were abandoned.

Progressive damage was a function of the decreasing distance to the caving line and the class of the rock mass. For example, in very good rock mass (competent quartzite), there was no damage for 10 months if the distance to the cave line was over 20 m. However, for the distance of 20 metres, the drive/crosscut suffered substantial damage in 30 months (Fig. 8.3). In good rock mass, some damage was visible within six months, although spalling was less than 0.5 m. In 12 months, the damage was substantial, and after two years, the excavations collapsed.

8.3.2 Case study of stress transfer through the down-dip abutment

An understanding of the geotechnical environment was critical to continue down-dip mining in a deep nickel mine. The orebody was mined by sublevel caving (SLC). Typically, the orebody dipped sub-vertical and was 80 m wide and 150 m long (Wood *et al.*, 2000). The hangingwall contact with ultramafic orebody was marked by a very prominent low-strength shear zone. The secondary excavations took place in a high-stress environment, in which the rock mass consisted mainly of ultramafic rock with uniaxial compressive strength varying from 50 MPa to 130 MPa (conditions varying from very poor to good).

Ore extraction by SLC was about 800 m below the surface, and production levels were 25 m apart (floor to floor) with ore crosscuts being 5 m by 5 m. Pillars between crosscuts were 6 m wide. In such conditions, mining-induced stresses were manifested in ground movement resulting in crosscut convergence ahead of production faces. The convergence took place horizontally. Figure 8.4 shows an example of the advanced stages of ground deterioration in a crosscut, with severe sidewall closure leading to support failure.

Figure 8.4 Horizontal closure of a crosscut (Szwedzicki *et al.*, 2007).

The very high rates of sidewall closure and floor heave led to the premature failure of ground support and in a few cases loss of access. In cases of lost crosscuts, the orebody was wrecked and recovered from the levels below. The adverse behaviour commenced prior to sublevel cave extraction (on the level) and increased progressively once caving continued (Wood *et al.*, 2000).

To determine the extent of fracturing, holes were drilled into the pillars was carried out using rod extensometers. It was determined that in competent ground, a zone of broken rock mass extended 0.5–2 m into the pillars, while in poor ground, the pillars between crosscuts were completely fractured.

In areas of poor ground, near and through the shear zone, crosscuts were prone to convergence, with horizontal closure and floor heave being common. Two major factors affecting the rate of closure were exposure time and proximity to the caving face. With stand-up time of 16 months or more, horizontal convergence monitoring showed that closure, started as abutment pressure, was transferred on the excavation on the down-dip levels. Initially, convergence accelerated up to 0.25 m per week with cumulative closure exceeding 2 m and with floor heave of up to 1.5 m (Szwedzicki *et al.*, 2007).

High rates of deformation resulted in the limited life of crosscuts, requiring that each excavation be carefully considered during the planning, development and rehabilitation stages. Planning and design included optimised timing of development (just-in-time) and consideration of mining geometry. Studies conducted on the optimisation of development and production called for the working life of the crosscuts to be between 16 and 24 months. Thus, planning and scheduling aimed at crosscut life of a maximum 24 months. When designing mining geometry, the following factors were considered: maximum open span of crosscuts and junctions, preferred T-junction to X junction, pillar width-to-height ratio, extraction ratio and the excavation shape. The controls were implemented during planning and design, by extraction sequencing (controlling abutment stress), using continuously improved ground support and crosscut rehabilitation.

8.3.3 Case study of stress transfer through compacted caved rocks

In a deep block caving mine, drawpoints in certain areas quickly deteriorated and had to be abandoned. The production level was 800 m below the surface and located in rocks with the uniaxial compressive strength of 130 MPa. The drawpoints were 4 m wide. Initially, three drawpoints in the middle of the panel squeezed and finally collapsed. Consequently, the number of closed drawpoints increased. The geotechnical assessment to the damage to the drawpoints indicated that:

- a high column of compacted caved ore resulted in stress that was transferred through the rock mass surrounding the drawpoints,
- compaction progressed vertically and horizontally, followed by progression of induced stress to surrounding drawpoints, and
- the load on drawpoints due to compaction increased in time.

An example of monitoring of a drawpoint closure in time is shown in Figure 8.5. For about 25 months, the drawpoint did not show any convergence. Once compaction took place, the drawpoint started to show some stress in the form of cracks in the shotcrete. All materials

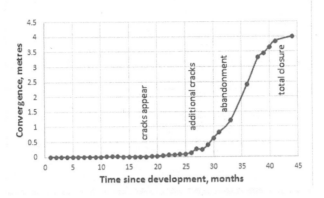

Figure 8.5 Results of monitoring of convergence of a drawpoint.

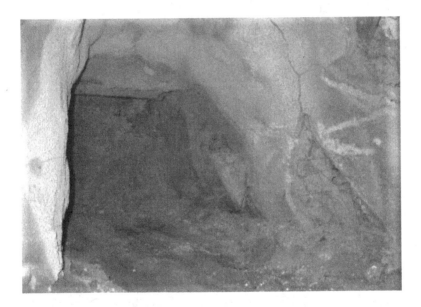

Figure 8.6 A drawpoint after 30 months.

inside the drawpoints were compacted, and stress was transferred to the surrounding areas. For the first six months of exposure time, the convergence was about 10 mm per month. That convergence did not affect mining production from these drawpoints. With closure increasing to 0.6 m, the drawpoints were closed for production and abandoned. With time, convergence started to increase exponentially from 50 mm per month to 100 mm per month, resulting in the final collapse. It could be concluded that although the life of the drawpoint was about 40 months, the exposure time to abutment stress was about 15 months.

Figures 8.6 and 8.7 show the progression of drawpoint closure in time.

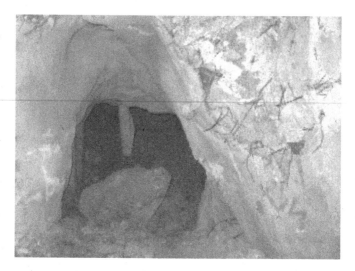

Figure 8.7 The drawpoint after 40 months.

Figure 8.8 Crack opening in a pillar between two drawpoints two days after draw was suspended.

A very fast deformation and closure of drawpoints and a drive was observed when the ore draw rate was limited and finally suspended. The original ore draw rate was about 160 tonnes/day. Due to high and large hang-ups, the draw rate was reduced to 44 tonnes/day. After one month, the drawing ceased due to rehabilitation work in the area. Almost imme-diately (within two days), the first signs of deterioration in the form of the crack opening were noticed in a pillar between two neighbouring drawpoints (Fig. 8.8). On the ninth day of ceasing drawing, stiff steel support showed deformation of about 200 mm. On the 14th day, vertical convergence was about 150 mm/day. Figures 8.9 and 8.10 show a progress in the closure of a ventilation rise in the area. In 10 weeks, the drawpoints and a part of the drive collapsed. It was estimated that about 5000 tonnes of rocks collapsed.

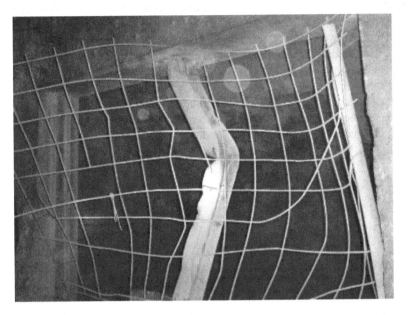

Figure 8.9 Stress-induced deformation in a ventilation rise nine days after suspension of draw.

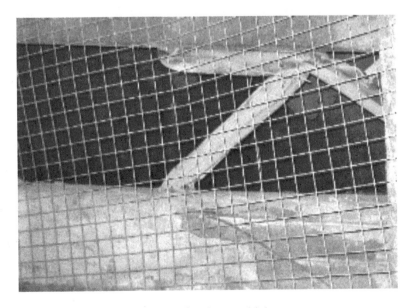

Figure 8.10 Closure of the ventilation rise within the next 14 days.

The following factors contributed to the deterioration of ground conditions: stress build-up due to a lower rate of draw for a period of 30 days (high hang-ups) with subsequent ceasing of draw, and existence of a shear zone/fault intersecting the access to the drawpoint. The damage that initially was highly localized expanded for about 55–60 m along the drive. During the first two weeks of the exposure time, vertical convergence accelerated up to 150 mm/day. However, after about 10 weeks (after large ground collapse), it subsided, with some residual convergence still taking place. Steel sets installed in the affected area started to deform after about two weeks after installation.

For three case studies at the mining depth of about 800 metres below the surface, exposure time, calculated from the first sign of stress transfer to closure of the excavations, depended on ground conditions, existence of structural features size of pillars, excessive open span and the distance to the caving line. The exposure time was as short as 10 weeks.

Planning for secondary excavations, in areas of expected stress transfer, should focus on a minimum distance to the excavation front and just-in-time development.

Chapter 9

Post-failure rock mass behaviour

After a large-scale failure, the rock mass usually doesn't stabilize immediately but exhibits some post-failure behaviour that can continue for a long time. An understanding of potential post-failure residual rock mass behaviour is critical when making decisions on re-entry time, continuation of mining activities and also long-term utilisation of the effected areas. In many cases of large-scale geotechnical events like disasters, rescue teams must enter effected areas. In such cases, there is a need to determine minimum re-entry time so that the teams are not exposed to hazards of further ground deterioration.

After large geotechnical events, the posed hazard is often high enough to warrant the exclusion or evacuation of personnel from mines. The period of exclusion until re-entry occurs is a decision for site geotechnical engineers and mine management that must balance the potential risk to personnel with lost production time and associated costs (Tierney and Morkel, 2017). There is currently no widely accepted method for determining re-entry times, and each hazard requires separate consideration. Mines often rely on decision-making based on rules of thumb, experience or common sense. Re-entry protocols could potentially reduce the risk to personnel from an early re-entry or reduce the lost production from an unnecessary exclusion. Re-entry protocols can only be expected to limit the exposure of personnel to large events.

Post-failure behaviour also must be considered when further mining activities are planned near damaged ground. This includes assessing the re-occurrence of the events and determining the size of protective and stabilising pillars and also to determine re-support requirements (Szwedzicki, 2005). Knowledge of long-term post-failure behaviour is critical for determining the long-term surface stability over closed underground mines. Progressive failure of the hangingwall or surface crown pillars may result in sinkhole formations and thus creating hazards to the general public.

Post-failure response of the rock mass is seldom observed, reported or recorded because there might be no immediate access to effected areas, geotechnical instrumentation might be damaged and the necessity of reporting such behaviour might not be understood. Some observed post-failure behaviours mentioned in reports of various commissions of inquiry are discussed below.

9.1 CASE STUDIES OF POST-FAILURE BEHAVIOUR

Post-failure or residual behaviours were reported after seismic events, gas emission, collapse or rockfalls, fracture formation, propagation of damage, subsidence and change in water inflow to mines.

Residual seismic response after a rockburst was monitored at Falconbridge mine, Ontario. Following the rockburst, seismic events were recorded by a microseismic monitoring system installed at the mine. In the first half hour after the rockburst, 116 microseismic events were counted. Two hours later, a series of large seismic events were recorded, and in the following 24 hours seismic activity faded away with only 10 significant events (Goodspeed *et al.*, 1995).

In the case of the pillar collapse at the Solvay Trona Mine, where a massive pillar failure occurred underground, it was recorded that ground falls continued for several hours thereafter. Additionally, increased emission of methane was noticed for a period of three months (Zipf and Swanson, 1999).

At Coalbrook Colliery, RSA, three collapses took place. For the three days after the first collapse, roof noises were heard around the perimeter of collapse and deterioration of the barrier pillars persisted but then died down. After the second collapse, 24 days later, strata movement continued over a number of days and extended to the area of subsidence. Cracks developing on the surface were noted, meaning that possible breakthrough to the surface occurred. The third collapse took place three hours later. Further falls took place in the workings and impeded the rescue attempts (Moerdyk, 1965).

In many cases, the residual response was reported after surface crown pillar collapses (see Chapter 3, Section 3.1). In most cases, the surface crown pillars were composed of weak rock or soil. When exposed to eroding, the unconsolidated material moved into exposed mine stopes. The examples are as follows:

> When a sinkhole developed over Chaffers Shaft area stopes, the sinkhole was immediately filled. However, the ground movement was observed to continue for about two years. The movement was noticeable as sinking of backfill material, crack opening and dislocations around the sinkhole.

> After Nobles Nob mine collapsed, circumferential cracks opened around the sinkhole throughout the next three years.

> Similar residual behaviour was observed at Warrego Mine. After the formation of the sinkhole, the surface continued cracking around the subsidence and ground movement was reported on the surface for the next few days. Cracking on the surface extended up to 80 m from the depression. Underground inspections revealed evidence of ground movement along the footwall contact and new cracks formed in various places. Two years after the collapse, through periodical ground movement, weathering and subsequent erosion of the oxidized material, the sinkhole had substantially increased in size. (Szwedzicki, 1999b)

A post-subsidence of ground around Scotia mine was periodically monitored. After the subsidence, the sinkhole was approximately 75 m long and 50 m wide (Fig. 9.1), the collapsed material being approximately 30 m below surface at the northern end and rilling down to about 50 m below surface at the southern end.

Surface ground movement instigated by the collapse continued as further subsidence in the form of gradual ground migration. That migration took place mainly during the rainy seasons. The surface, two years after the collapse, is shown in Figure 9.2.

Twenty years later, the diameter of the sinkhole had increased from 70 m to more than 200 m (Figure 9.3); this can be attributed to weathering and the deterioration of the oxidized zone in the upper part of the sinkhole. The average depth from the surface to the bottom of the

Figure 9.1 A sinkhole developed above the Scotia Mine, 1974 (Szwedzicki, 1999a).

Figure 9.2 A sinkhole above the Scotia Mine two years after crown pillar collapse (Szwedzicki, 1999a).

Figure 9.3 Sinkhole above the Scotia mine 20 years after crown pillar collapse (Szwedzicki, 1999a).

sinkhole had decreased from 45 m to 30 m. The estimated volume of the collapsed material in 1974 was about 200,000 m³ of rock. Calculation of the volume of the sinkhole 20 years later showed that the volume of migrated material was 500,000 m³ of rock. The conclusion was that the rock mass was progressively weathering and migrating, and that in 20 years about 300,000 m³ of additional material had moved into the mine openings (Szwedzicki, 1999b).

Following the failure at Telfer Mine, monitoring continued. Instrumentation outside the failure area showed a dramatic decrease in the rate of movement. Within a few months, several cracks showed negative movements corresponding to cracks closing behind the crest of the slope.

Re-occurrences of the disastrous events were recorded at Lassing Talc Mine, Austria, Coalbrook Mine, Warrego Mine, Perseverance Shaft and Mufulira Mine.

- At Lassing Talc Mine, water and mud entered underground excavation. About nine hours later, when a rescue team entered the mine, a second inrush took place.
- At Coalbrook Colliery, the barrier pillars that temporarily arrested the movement of the strata were slowly failing over a period of a month, culminating in the second and larger collapse.
- At Warrego Mine, the third and fourth subsidence phases took place 9 and 12 months later.
- A long time span between first discontinued subsidence and chimney formation was reported at the Perseverance Shaft area; after the initial chimney formation, the mining activities continued and 16 years later the second subsidence developed (in exactly the same place).

Case studies of ground collapse revealed that major collapses were followed by seismic activities in the form of rock noises, rockfalls and pillar fracturing, which took place for a period of a few hours, if not days. In cases of crown pillar collapse, some slumping of weathered rocks and soil material from the walls of sinkholes was noticed. Ground movement and additional subsidence have often been noticed for several years, usually accelerated after heavy rain.

9.2 THE EFFECT OF MINING GEOMETRY ON POST-FAILURE BEHAVIOUR

The post-failure behaviour of the rock mass is affected by mining geometry. It can be compared to post-failure behaviour of rock samples of different geometry tested in a laboratory (see Chapter 2).

With increasing pillar width:height ratio, post-failure behaviour changes from brittle to ductile. Brittle fracture is said to occur when sudden loss of strength follows limited plastic deformation. Ductile deformation or strain hardening is said to occur when rock can sustain further permanent deformation without losing load-carrying capacity. For slender pillars the behaviour is typically brittle; for squat pillars it is typically ductile. (Except at the pillar perimeter, where brittle failure is typically observed.) Example of stress-strain curves of various width:height ratios for rock samples for slender, regular and squat model pillars are shown in Figure 9.4.

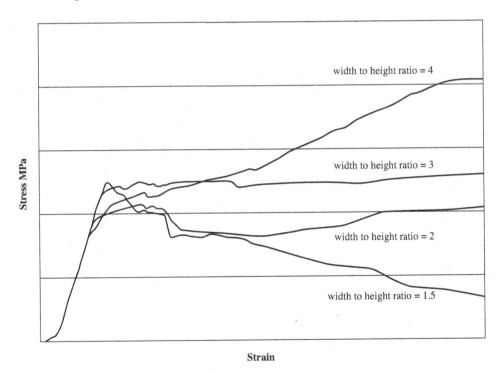

Figure 9.4 Post-failure behaviour of rock samples of different geometry (Szwedzicki, 2000).

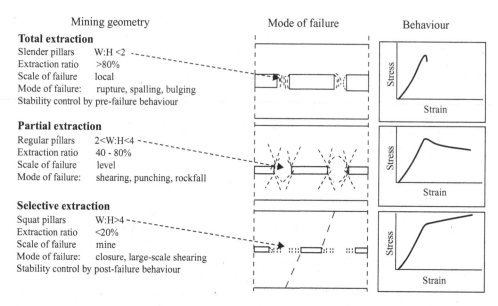

Figure 9.5 Generalised rock mas behaviour in a high-stress environment for various mining geometry (Szwedzicki, 2000).

The peak strength did not substantially alter with the change in the width:height ratio, and some variations can be attributed to mineralogy and microscopic structural features in rock samples. However, post-failure strength increased substantially, and for the width: height ratio of 4 (at strain of 0.04), the post-failure strength was almost twice as high as the peak strength.

Generalised rock mass behaviour showing relation between mining geometry, mode of failure and behaviour in a high-stress environment is summarised in Figure 9.5 (Szwedzicki, 2000).

In selective extraction (i.e. squat pillars and low extraction ratio) in a high-stress environment, mining failures are controlled by structural features rather than by local stress concentrations. Although local stress concentrations can be observed as pillar hourglassing or fretting (small rock fragments), movement along shear zones, closure of openings and rock mass deformation takes place in the hangingwall and/or footwall. Shear movement can extend for a long distance. The pillar deformation is determined by post-failure ductile behaviour and post-failure strength of squat pillars is a function of confinement caused by fractured rock.

In partial extraction (i.e. pillars – width:height ratio between 2 and 4 – and extraction ratio between 40% and 80%), rock and rock mass failure can be affected by stress and structure. Failure is manifested as pillars punching the hangingwall and footwall, often resulting in roof guttering and floor heaving. After reaching the peak strength, samples progressively fail, exhibiting some post-failure strength with formation of stress-induced circular fractures. Pillars exhibit rock spalling at the sidewalls of the sample and fracturing and punching of the hangingwall and footwall parts of the sample. A limited ductile post-failure behaviour was observed.

In high-extraction conditions (i.e. slender pillars and high extraction ratio), it is the local rock failure that is predominant. Under high stress, pillars rupture, spall, slab or shear, and rock fragments are relatively large. Although the failure may be influenced by local structures, the failure is stress controlled, i.e. it takes place due to the high concentration of mining-induced stress in the pillars. Usually the hangingwall or the footwall is not affected directly by stress. Pillar stability is determined by the pre-failure properties of rock, and once the pillar fails it does not transfer any stress. It should be noted that failure of a number of slender pillars can lead to whole mine (or part of the mine) instability.

Increasing the pillar width:height ratio or lowering the extraction ratio does not necessarily prevent rock mass failure. The failure can occur at any width:height ratio or extraction ratio if sufficient load is provided. The advantage of a pillar design with high width:height ratio is that such pillars exhibit high post-failure strength and behave in a ductile manner. Pillars that are slender exhibit brittle behaviour.

The factors affecting mine and local stability should be considered when designing a mine.

Chapter 10

Modes of failure of rock and rock mass

Mode of rock mass failure is defined as a manner, form or mechanism of rock or rock mass fracturing leading to failure under induced stress. The location, orientation, size, density and extent of the discontinuities contribute to different modes of failure, and these affect the mechanical parameters of rock samples and rock mass.

Modes of failure around underground excavations are functions of multiaxial stress conditions, the existence of geological structures and the geometry of mining excavations.

Although the process of rock mass failure is a function of rock properties, structural features and mining geometry, the failure itself is driven by stress changes. Depending on the stress path, the process can be progressive or transgressive.

During a progressive failure process, the rock mass deteriorates with an increase in stress. After reaching stress balance, the failure process ceases, and the rock mass stabilises. For such behaviour, point of failure is not easily determined. Examples of such failure are floor heave, convergence or progressive slope movement.

During transgressive failure process, with an increase stress, the rock mass reaches its peak strength and fails with a clearly recognisable brunt. At the failure point, disintegration can be gradual or violent. Examples of gradual process of failure are slabbing, disintegration or fracturing, and slope failure. Violent failure can take place in the form of collapse, outburst or rockburst.

Each mode of failure can be described by the location (in reference to mining geometry), extent of failure and duration of the failure process.

On a macroscopic scale, the discontinuities are characterised by distinctive joints, and the tensile strength of rock mass is often considered to be zero. At the other end of the geotechnical domain scale, there are microscopic imperfections like micro-defects, intergranular cracks or micro-flaws.

At a microscopic scale, i.e. the scale of rock samples, the detection of discontinuities and their effect on the mechanical properties of rocks was carried out using a non-destructive testing method. Although such discontinuities are sometimes visible, many of them remain undetected during laboratory sample preparation and testing. Microscopic rock samples that include discontinuities can be considered to be a form of transition between the intact rock material and the rock mass.

10.1   MODES OF FAILURE AT A SAMPLE SCALE

Rock samples contain randomly oriented discontinuities. Under increasing uniaxial compression, stress in rock samples redistributes around such cracks. Sample failure is the result of micro fracturing, the nucleation of cracks (tensile fractures of the microscopic scale) at points of stress concentration and their propagation along the direction of the maximum principal stress. Rock samples under uniaxial compressive stress, due to localised stress concentrations around microscopic discontinuities, can fail in tension, in shear or in coupling of the tension and shear stresses. The complete set of relations between the strain and stress components can be written in a matrix form – known as the compliance matrix. The architecture of the elastic compliance matrix was illustrated conceptually by Hudson and Harrison (1997) and is shown in Figure 10.1.

Two main categories of brittle fracture for a rock specimen subjected to uniaxial compression – axial cleavage and conjugate shear – were identified (Gramberg, 1989). It was widely postulated that a mode of failure depended upon the degree of the end constraints of the sample offered by the platens of the testing machine and the surface quality of the parallel ends of the sample.

The way the sample fails is reflected by the mode of the sample failure. This means that the mode of failure affects the resultant strength of the sample. Analysis of the mode of failure provides insights into the orientation of principal stress in rock samples. Extension fractures develop at right angles to the minimum principal compressive stress direction and will contain the orientation of the maximum and intermediate principal compressive stress.

The existence of microscopic discontinuities is responsible for "scale effects", i.e. strength reduction with increased sample size. The larger the sample the higher the probability that the discontinuities will affect the strength and that the sample will fail in a shear mode.

It is hypothesised that modes of failure can be classified similarly to the compliance matrix reflecting various stress conditions (Fig. 10.2). However, it should be noted that the theory of elasticity does not apply to failure, and there is no physical link between the elastic compliance matrix and the matrix of failure modes.

It can be argued that the value of compressive strength obtained when the rock failed in shear mode represented properties along the discontinuities, whereas the value obtained when the rock failed in extension mode represented the strength of intact rock material.

Examples of cylindrical rock samples that failed in various modes were grouped in the matrix of modes of failure and are presented in Figure 10.3.

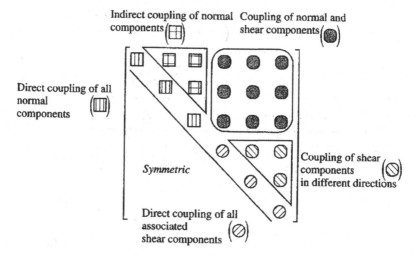

Figure 10.1 The architecture of the elastic compliance matrix (Hudson and Harrison, 1997).

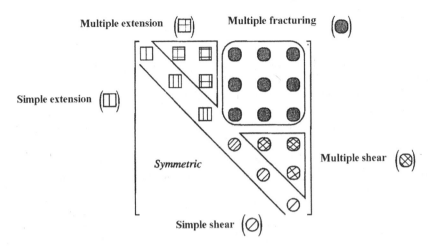

Multiple extension (⊞) Multiple fracturing (⊙)

Simple extension (⊡)

Symmetric

Multiple shear (⊗)

Simple shear (⊘)

Figure 10.2 The architecture of a matrix of failure modes of cylindrical rock samples tested in uniaxial compression (Szwedzicki, 2007b).

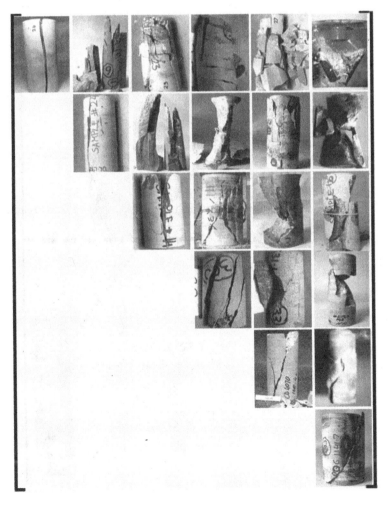

Figure 10.3 A matrix form of failure modes of rock samples tested in uniaxial compression (Szwedzicki, 2007b).

It should be noted that during testing of rock samples from a single geological formation, the mode of failure is usually similar for all samples due to the same type and orientation of defects that contribute to the same mode of failure.

The results of uniaxial compressive strength (UCS) tests on rock samples are open for misinterpretation. The results of the UCS of rock samples show the coefficient of variation considerably higher than the results of the UCS of manufactured homogenous materials.

It was hypothesised that the results of tests on rock samples are not strictly comparable, because tensile and shear components may vary in magnitude depending on the mode of failure. The reason for this is principally due to discontinuities that affect the modes of failure.

It is asserted that the variations in values of uniaxial compressive strength of samples from the same lithology depends on the mode of failure. With the same mode of failure, variations may be relatively small. However, when various modes of failure take place on similar samples, the variations can often reach extreme values. As an example, four UCS tests on

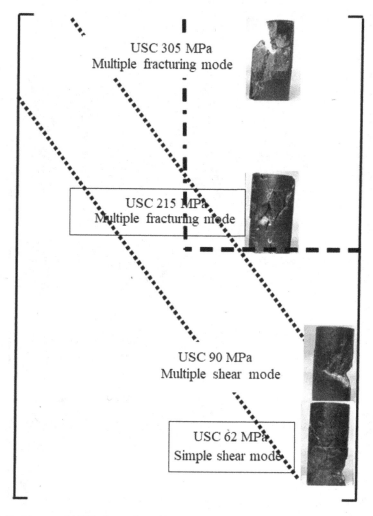

USC 305 MPa
Multiple fracturing mode

USC 215 MPa
Multiple fracturing mode

USC 90 MPa
Multiple shear mode

USC 62 MPa
Simple shear mode

Figure 10.4 Positions and UCS of samples of basalt that failed in various modes in the matric form (Szwedzicki, 2007b).

cylindrical contiguous samples of basalt that failed in different modes are presented. The sample that failed in the simple shear mode had the UCS of 62 MPa. The sample that failed in the multiple shear mode had the UCS of 90 MPa. Two samples failed in the multiple fracturing mode. One sample that failed predominantly in shear stress had the UCS of 215 MPa and the other that failed predominantly in tensile stress had the UCS of 305 MPa. The value of the UCS and the positions of the samples in the mode matrix are shown in Figure 10.4.

It appears that different modes of failure are due to the microscopic discontinuities in rock samples rather than variations in sample preparation, test procedure or end-boundary conditions.

Analysis of the modes of failure makes it possible to determine whether the rock samples failed in tension, in shear or in coupling of tension and shear.

The way the sample fails (i.e. mode of failure) affects the obtained strength of rock samples. The value of the UCS is a function of the mode of failure – the values of the UCS obtained in extension are the highest, whilst the values obtained in shear are the lowest.

Understanding the effect of the mode of failure on strength enables an in-depth interpretation of the results of the uniaxial compressive strength tests.

10.2   MODE OF FAILURE AT A LOCAL SCALE

Observations of the mode of failure of rock samples tested in laboratory conditions in uniaxial compression can be extended to modes of failure of square pillars in room-and-pillar mining (Brady and Brown, 1993). Modes of rock mass failure are determined by multiaxial stress conditions, properties and orientation of geological structures and mining geometry. The geometry of the excavations, within the zone of influence, often has a dominant effect on the stress concentration around underground excavations.

The rock mass around underground excavations is subjected to unique stress configuration, i.e. low confinement and the failure modes. The modes of failure are structurally controlled or stress induced but most often controlled by a combination of them. Structurally controlled failures are most frequently observed at shallow depths, while stress-induced failure is found in deep mining. Modes of failure can be grouped by their controls, severity, numbered of unconfined surfaces, progression and duration of failure, production and development areas, etc.

The majority of structural rock mass failures are observed in weak rock mass and in areas of large open span, i.e. excavation intersection. The majority of stress-induced failures are observed near stress abutments or in pillars that have up to four unconfined surfaces. A variety of failure modes are recorded in pillars. In the rock mass defined by pillar faces, confining forces are zero or very low and the mining geometry plays a major role in defining the mode of failure. Examples of various modes of failure, taking place in a variety of mining situations, are provided below.

Single extension failure is shown in Figure 10.5. The tensile crack opened perpendicular to the unconfined face formed by a sidewall of a drawpoint access.

Multiple extension failures took place through a shaft pillar in a gold mine at a depth of 400 m. With an increase in extraction ratio, the load on the pillar increased. The pillar slabbed parallel to the maximum principal stress and that resulted in pillar dilation (Fig. 10.6).

Shearing and bulging modes are often encountered on pillars between crosscuts. To optimise draw of ore, such pillars are usually not wide enough to support the hangingwall. Figure 10.7 shows a damaged pillar in a gold mine (about 400 m below the surface), while Figure 10.8 shows a pillar in a mine (about 800 m below the surface).

Under high stress, a pillar fractured at mid-height, forming an hourglass shape (Fig. 10.9). Despite a reduction in supporting area, the fracturing ceased and there was no progression in damage.

In a colliery, using board and pillar mining, the coal seam being stronger than the roof caused roof guttering (Fig. 10.10).

Figure 10.5 Single crack opening in a drawpoint access.

Figure 10.6 Vertical splitting of a shaft pillar (Szwedzicki, 2003b).

Figure 10.7 Shearing and bulging of a sidewall between crosscuts (Szwedzicki, 2003b).

Figure 10.8 Bulging and shearing of a crosscut sidewall.

Figure 10.9 Pillar hourglassing (Szwedzicki, 2005).

Figure 10.10 Pillar punching the roof (Szwedzicki, 2003b).

In a zinc mine, due to the relaxation of the rock mass, the clamping forces ceased to confine the rock mass and a rockfall took place. The relaxation took place after the mining front of the orebody progressed below the position of the drive.

Structurally controlled fall of ground took place in a crosscut (Fig. 10.12). The unconfined surface at the back of a crosscut allowed failure along intersecting planes of weakness.

Figure 10.11 Fall of ground due to stress relaxation in a zinc mine.

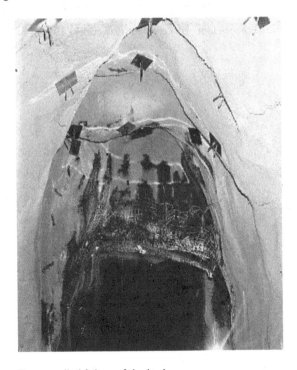

Figure 10.12 Structurally controlled failure of the back.

A crosscut in a weak rock mass was heavily supported by bolts, wire mesh and shotcrete. When it came under stress, a shear failure initiated in the shoulder took place along the crosscut (Fig. 10.13).

A crosscut in a sublevel caving mine was damaged by floor heave and sidewall convergence. Roof sagging was negligible (Fig. 10.14).

Figure 10.13 Shear failure at the shoulder of a crosscut (Szwedzicki, 1989a).

Figure 10.14 Floor heave and horizontal convergence in a crosscut.

A horizontal closure and floor heaving on one side of the crosscut resulted in asymmetrical damage leading to abandonment of a crosscut, as shown in Figure 10.15. The asymmetry in closure resulted from different width of pillars on both sides of the crosscut. Larger convergence and higher floor heave was noticed at the side where the pillar was slender.

In a high-stress area within a copper mine at depth of 700 m, a crosscut was heavily supported with concrete walls and stiff steel support. Figure 10.16 illustrates the damage to supported crosscut during a rockburst.

Figure 10.15 Asymmetrical closure of a crosscut due to approaching caving front.

Figure 10.16 Damage to steel and concrete support due to rockburst (Szwedzicki, 1989a).

10.3   MODE OF FAILURE AT A MINE SCALE

Uncontrolled ground movement in the form of collapse of rock mass often results in discontinuous subsidence. The most common collapse modes are sharply defined cylindrical depressions or shaft-like holes reaching the ground surface (Goel and Page, 1982), commonly known as sinkholes.

The process of collapse is initiated underground by the fall of unstable blocks (when structural planes of weakness are exposed in the back of the excavation) or by the failure of intact rock. Both of these mechanisms depend on a critical dimension of the exposed excavation area at which caving will progress. Progressive deterioration of a hangingwall and rockfalls above mined-out stopes in steeply dipping orebodies can result in the progression of created voids to the surface.

In subsidence engineering, it is well known that the dimensions of extraction panels should meet certain critical values before full surface subsidence develops. The potential collapse is a function of the inclination of the orebody – in steep orebodies, the caved-in material rills down and may not provide support to the hangingwall along the whole length of its dip.

The rock mass around excavated stopes can deteriorate, which may ultimately lead to rock mass collapse. The process of ground collapse can be considered as taking place in four stages (Fig. 10.17):

*   In the first stage (excavation scale), which could happen years before the failure, local stress changes take place near the mined-out stopes. It may take place in a form of change in ground conditions such as movement across shear zone, local instability in drives or uncontrolled ground movement.
*   In the second stage (level scale), which is evident one to three months prior to the failure, progressive deterioration of the stopes hangingwall takes place. Small rockfalls affect the continuity of the hangingwall. Isolated blocks of rock rill to the bottom of the stopes and do not provide any support.
*   During the third stage (level scale), one to four weeks prior to collapse, pillars between mining levels deteriorate to the point of failure. Failure can be caused by reduction in strength properties of material due to water and weathering, seismic events, or blasting carried out nearby. Broken rock is rilling down to the bottom of stopes and a large unsupported open span is created.
*   In the fourth stage (mine scale), a large unsupported span exceeds its critical length and starts to cave. Within days or even hours before collapse, rock noises and rockfalls are recorded in various places of the mine. The failure is usually progressive through strata and collapsed material rills down the stopes. The filling process prevents self-support of the hangingwall by broken material, which increases its volume during the process of fragmentation.

The statistical distribution of recorded pre-failure rock mass behaviour is shown in Figure 10.18. The distribution shows that not all stages are recorded in each case study. Pre-failure rock mass behaviour calls for the proper selection of geotechnical monitoring instruments and systems. During the first stage, during which the hangingwall starts to deteriorate, the most appropriate monitoring method is to carry out regular inspections to record the signs of the hangingwall deterioration. The deterioration can be observed as fall of blocks, crack opening, shear movement across the geological formations or the hangingwall sagging. The

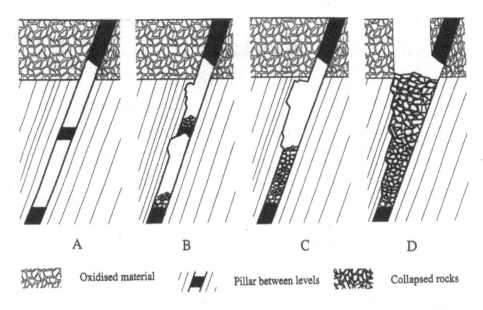

A   B   C   D

Oxidised material   Pillar between levels   Collapsed rocks

Figure 10.17 Progression of scale for mine ground collapse – (A) initial state of mining stopes; (B) deterioration of a hangingwall; (C) progressive failure; (D) rock mass collapse (Szwedzicki, 1999a).

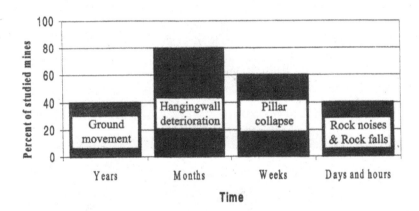

Figure 10.18 Type of ground response prior to failure (Szwedzicki, 1999a).

geotechnical instrumentation that can be applied is borehole extensometers, crack monitoring plates or load cells on cable bolts. In the second stage, when the collapse of pillars between levels is expected, the most suitable is cavity monitoring. In the third stage, when the failure is imminent, it is possible to monitor seismic events. This can be done by recording the energy of each event or even by counting the number of seismic events. The results of ground movement recorded by extensometers, cavity monitoring and seismic events must be interpreted concurrently. The post-failure behaviour is noticed in a form of rock noises and slumping of the material on the surface. Ground movement and additional subsidence was noticed for many years, usually accelerated after heavy rains.

The size of unsupported span can increase with time; this can be caused by progressive pillar collapse or by planned pillar recovery. Often, larger, high-grade pillars are recovered in the later stages of mining and the hangingwall is left not adequately supported; the remaining pillars may be too small or too widely spaced to provide the required permanent support.

Following collapse of the final arch of ground, a depression is manifested on the surface. The failure of intact rock or the falling of unstable blocks is accompanied by an increase in volume of the failed material. The vertical distance to which caving extends into the hangingwall depends on the bulking of the material and the volume of the mined-out deposit. The natural bulking of the caved rock increases the volume of broken rock by a factor of 1.1–1.5 and can result in the development of self-support in the void, such that the upward collapse process is halted. Because of this, although the potential for cave propagation is high, usually only a few reach the surface.

Immediately after formation, the walls of the depressions are steep, but they may deteriorate with time through weathering or change in the mechanical properties of the ground. This may create a potentially unstable edge zone, near which increased lateral movement occurs by erosion of the sides or slumping of weak material. In general, if large voids are created near the surface, extensive caving is to be expected and risk of collapse is apparent. With smaller, deeper voids, the risk of caving progressing to the surface is smaller.

The surface reflection of caving can be defined by three distinctive zones: the caved-in zone, the fractured zone and the stable zone. Each of these zones displays a characteristic mode of failure, as described by the following: (Tyler et al., 2004)

- Caved-in zone – defined in terms of absolute movement and rates of movements. The interface with the fracture zone represents a discontinuity.
- Fractured zone – cracks up to 50 mm wide are evident. Figure 10.19 shows the fractured surface above the cave zone.
- Onset of the stable zone – hairline joints to visible cracks.

Figure 10.19 A fractured surface above a caved-in zone.

Chapter 11

Behaviour of fragmented ore

Behaviour of fragmented ore and waste rock can affect the solid rock mass response to mining. Rock mass fragmented after blasting usually compacts and that may not only provide confinement to the surrounding rock mass but can allow for transfer of stresses. Such transfer may result in ground deterioration around drawpoints and crosscuts on the extraction level.

Movement of fragmented ore and waste rock by draw affects rock mass response in caving zones, open stopes and orepasses. Draw enhances caving and prevents broken ore compaction. Slow draw rate or suspension can prohibit caving, causing structural damage to drawpoints and water accumulation (that can cause mud rushes). Controlled and periodical draw is used to control wall deterioration/collapse in the stopes (e.g. shrinkage stoping) or orepasses. In orepasses, it is a common practice to keep orepasses full, and limited draw is used to prevent wall deterioration by wall sloughing and formation of breakouts. There is also a direct relationship between rock mass response and fragmented material movement. Geotechnical instrumentation by stress measurement and convergence monitoring installed near stopes and orepasses show reduction in stress and convergence rate as soon as draw rate is increased (Szwedzicki, 2007a).

The behaviour of fragmented ore (ore flow) in the cave zone in response to drawing is governed by ore fragmentation, rock size distribution and drawing practices. It is also indicative of damage to the surrounding rock mass. Through evaluating the size distribution of rock fragments at the drawpoints, it is possible to determine whether the size distribution is optimal and whether blasting is effective.

In cave mining, width of the draw (extraction) zone is a fundamental mine design parameter and is used to determine the distance between drawpoints. For block caving, the optimal drawpoint spacing is where the draw zones of neighbouring drawpoints just overlap or touch each other. The size and shape of a draw zone is a function of design geometry (the drawpoint size and spacing) and the characteristics of broken rock (rock mechanical properties, fragmentation, material compaction and moisture).

With caving, a vertical cross-section of the draw zone is commonly assumed to be ellipsoidal. However, it has been noted that the draw zone can take the shape of a pipe or cylinder or can be irregular and may change in time with drawing (Brown, 2007). The development of the draw zone depends on the size distribution and packing of the material. High packing (compaction) results in the development of narrow extraction zones (Rustan, 2000).

Fragmentation and rock size distribution at the drawpoints determine migration and thus affect operational parameters, including the mining layout (spacing between drawpoints), the selection of mining equipment (like Load-Haul-Dump equipment and requirement for rock breakers) and need for secondary blasting. Fragmentation in a cave zone is directly linked to rock mass conditions. The finer the fragmentation of the rock, the closer the draw-point spacing should be. Guidelines to determine drawpoint spacing and extraction zone geometry were published by Laubscher (1995) and by Hustrulid (1999).

Two case studies of ore flow in caving mines and the behaviour of broken material in orepasses are presented.

11.1 ORE FLOW – A CASE STUDY FROM A BLOCK CAVING MINE

On the extraction level of the Deep Ore Zone (DOZ) Mine, PT Freeport, Indonesia, a tyre marker and tramp material, such as support elements, drilling accessories and a steel feeder, reported to the drawpoints. The marker and the tramp material migrated from the Intermediate Ore Zone (IOZ) block caving mine, which was located 320 m above. The material started to report after a draw of 20,000 to 150,000 tonnes.

The DOZ block cave was designed to be more than 1000 m long and more than 500 m wide. The mine had 857 drawpoints. Panel drifts were 4 × 4.4 m with the width of the drawpoints being 3.6 × 3 m. The drawpoints were on 18 m centres along the panel drift, resulting in column footprint of 15 × 18 m (Sahupala and Srikant, 2007).

A number of specific geological units were encountered in the DOZ Mine. A marker and tramp material were found in diorite and magnetite-fosterite, fosterite skarn, dolomite-marble and highly altered localised ore (HALO):

- Diorite and fosterite rock mass form a hard, competent rock with very good ground conditions.
- Magnetite skarn is a massive unit of competent rock with fair ground conditions.
- Dolomite-marble and HALO rock have highly variable properties and local conditions ranging from poor to very poor.

When the IOZ mine was closed, there were some steel items left on the extraction level (such as steel sets at drawpoints and panels, feeders to ore bins and some diamond drilling accessories). Those items were not salvageable or not salvaged because of deterioration that happened quickly. These items were expected to be recovered at the DOZ drawpoints. About two and a half years after abandonment, immediately below the IOZ Mine where the DOZ Mine caving operations took place, the first tramp material from the IOZ Mine was noticed. Subsequently, in the following three years, in 28 drawpoints tramp material was reported 55 times. During closure operations of the IOZ Mine, tyre markers were placed in selected locations on the extraction level. The tyres were from underground loaders (of 1.6 m diameter). Each tyre had a unique number burnt in and its exact location was recorded on mining plans. For the first tyre marker recovered, the horizontal movement was calculated to be 2–3 m only. Over 320.5 m vertical distance, the horizontal movement could be considered negligible.

The tyre maker was recovered two years and three months after drawing operations started from a drawpoint below.

The following tramp material, mainly used as ground support elements at the IOZ Mine, was found, including pipes, drilling rods, steel beams, wire mesh and bolts, and timber (Figs. 11.1 and 11.2). In most cases, they were steel beams. The steel beams were 3 m long with their H-cross section being 0.2 × 0.2 m.

Figure 11.1 A drilling rod in a vertical position (Sahupala *et al.*, 2010).

Figure 11.2 Steel beam found in a drawpoint (Sahupala *et al.*, 2010).

The original location of service pipes, drilling rod, etc., on the IOZ extraction level was not known, so the horizontal movement could not be determined.

An interesting case study was that of a steel feeder left on the IOZ extraction level. The feeder, 3 × 2 × 1.5 m, was found in "relatively good condition" in drawpoint on the DOZ extraction level. The feeder was recovered in the same drawpoint as the above-mentioned tyre marker. As it was a consecutive report of the tramp material in the draw-point, the appearance of the feeder was not used in calculations of the diameter of the draw zone.

Analysis revealed that the feeder migrated horizontally for as much as 126 m to the south (over 324 m vertical movement). Possible explanations of such extreme horizontal move-ment could be attributed to rilling of the caved material (and the feeder) as the subsequent caving events took place in the direction of the progressive cave.

By estimating the time from the moment the DOZ cave intersected the location of a tyre marker and the feeder on the IOZ level and the distance between the IOZ and DOZ extrac-tion levels, it was possible to calculate the average daily migration. The 1.6 m diameter tyre average migration was estimated to be about 0.37 m. On assumption that the feeder rilled down (at angle of 45°) from its original position to the same position as the tyre marker, it was estimated that the daily migration of the 3 × 3 × 5 m feeder was 0.15 m, i.e. two and half times slower.

11.1.1   Rock material in a draw zone

Over 320 m between the DOZ and IOZ extraction levels, the ore drawn in each zone inter-sected various geological/geotechnical formations. In each drawpoint, over the life of the mine, at least four different types of rock were drawn. However, for classification, the mate-rial that was drawn together with the tramp material was assumed dominant.

To calculate the diameter of the draw zone, the following assumptions were made:

- The shape of the ore draw zone was cylindrical.
- Rock fragments movement/migration takes place as mass movement, i.e. all fragments within the draw zone move at the same rate.
- Tramp material flows at the same rate as rock fragments.

Sand model studies show that material in the centre, i.e. immediately above the draw area, moves at a faster rate than fragmented material at the periphery of a draw zone (extending towards the neighbouring draw zones) (Hustrulid and Kvapil, 2008). Should the material move slower at the periphery of the draw zone, the calculated diameter of the draw zone could be smaller than the factual diameter.

In a number of cases, several items of tramp material were reported in the same draw-points. It was noted that many long and narrow elements like steel beams or timber were found at the drawpoints in a vertical position, suggesting that the material moved in the centre of the draw zone.

In the draw zone, the rate of movement of fragments depends, among other things, on the size distribution and the shape of fragments. When fragments are of various sizes, the smaller fragments move in between the larger ones and move faster. Fragments that are cubical or circular move faster than angular fragments or fragments of a different aspect ratio. Tramp material from the IOZ Mine found at the DOZ extraction level varied in shape and size. The smallest and probably cubical shape was that of a reamer, with the most elongated being steel beams, most of them up to 4 m long and 0.2 × 0.2 m wide. While it can be assumed that the reamer was moving at the same rate as the fragmented material, it is reasonable to assume that the remaining tramp material moved slower than the rock fragments. Should the material move slower, a larger volume of rock would be drawn before the tramp material reports at the extraction level. That is why the calculated diameter of the draw zone could be smaller than the actual diameter. However, for calculation of the draw zone diameter, it was assumed that fragmented rocks and tramp material moved at the same flow rate.

11.1.2   Diameter of a draw zone

The diameter was calculated by dividing the volume of the material drawn before the tramp material reported to the drawpoint below. The relation between the tonnage drawn and volume was determined using the *in situ* density of the various rocks and the swell factor of the caved rock.

The swell factor is a characteristic parameter of a rock mass and describes the volume change that occurs when the *in situ* rock mass is fragmented. The swell factor determines the *in situ* density of the caved rock and depends on fragmentation (uniformity and size), hardness of the rock, moisture content and compaction due to the height of the cave column. The swell factor is a parameter difficult to determine, and it may vary due to compaction and may change in time as the cave matures. Based on the observations, a swell factor of 120% was used for calculation of the draw zone diameter.

In several drawpoints, the tramp material was reported a number of times. Although it was possible that the subsequent reports of the tramp material were due to different shape and orientation of the material during the flow, it also could be inferred that the subsequent tramp material was not immediately over or in the centre of the considered draw zone. For this analysis, only the first reports of the tramp material were used in the calculations (Sahupala *et al.*, 2010).

For all case studies, the diameter was in a range from 6.28–14.59 m with an average of about 12 m. With a drawpoint width of 3.6 m, this means that the diameter was about 8 m wider than drawpoints. It should be noted that the smallest diameter of 6.28 m was calculated for a small size drill reamer.

It was noticed that irrespective of various geological origins of ore, i.e. various ground conditions, the diameter of the draw zones was found to be of similar range.

An interesting relationship was observed by plotting the relation between the diameter of the draw zone and the tonnage drawn (Fig. 11.3). For a range of draw from 23,000–50,000 tonnes, there was a linear relationship with the diameter which ranged from 6–15 m. This indicates that the larger tonnage drawn, the larger the diameter. It can be concluded that the diameter of the draw zone, within the same draw height of 320 m, was dynamic, i.e. increased with the tonnage drawn.

An analysis of the draw rate for 28 drawpoints was carried out. The rate of draw from good to poor ground was in a range from 100–160 t/day. Analysis for draw between 100–160 t/day indicates that the rate of draw does not affect (or affects to a minimal extent) the diameter of the draw zone (Fig. 11.4). In two case studies, for the draw rate in a range of 200 t/day, the graph shows some increase in the diameter. However, this could be a statistical variation and more data is needed to confirm that observation.

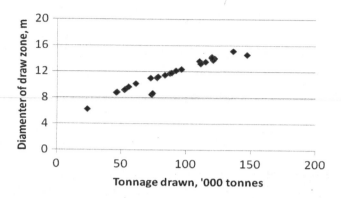

Figure 11.3 Increase in diameter of the draw zone with increased tonnage drawn from the drawpoints (Sahupala et al., 2010).

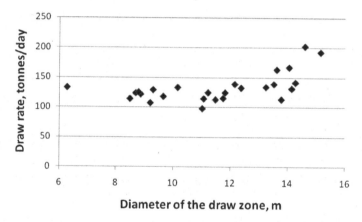

Figure 11.4 Diameter of the draw zone versus draw rate (Sahupala et al., 2010).

11.1.3 Change in fragmentation due to draw

An analysis was conducted to determine how fragmentation changes with the tonnage drawn, how the fragmentation affects the diameter of the draw zone, and whether the fragmentation affects the rate of migration of fragments and the tramp material from the IOZ to the DOZ Mine. Although it was evident that the fragmentation decreases overall as more tonnes are drawn, the variations in fragment size made statistical analysis unacceptable. The main reason was that large and oversized fragments continued to report occasionally to the drawpoints over their life. However, it was visually assessed that at the beginning of the caving in very good ground conditions, the median fragment size was 0.2–0.5 m (Fig.11.5). With material drawn to 100,000 tonnes, the median fragment size was generally in the range of 0.02–0.07 m (Fig.11.6). In cases where tonnage exceeded 150,000 tonnes, the median fragment size was less than 0.01 m with a relatively large portion of fragments less than 0.001 m. From observations of the drawpoints, it was evident that very fine and fine material was flowing in between large and oversized blocks.

Figure 11.5 Fragmentation in early stage of draw (Sahupala *et al.*, 2010).

Figure 11.6 Fragmentation from a mature cave (Sahupala *et al.*, 2010).

11.1.4   Effect of ore draw rate on damage of drawpoints

In high stress areas, drawpoints can suffer substantial damage in which includes damage to support, floor heave and destruction of concrete walls and back. It was observed that limiting of oredraw from the drawpoints increases convergence rate. Increased convergence rates was caused by compaction of fragmented ore in the drawpoints that would restrict flow of ore and increase load on the pillars. The daily draw was 66–88 tonnes per shift. To keep the area open, the draw rate was increased and guidelines were developed (Sahupala and Srikant, 2007). The criteria for the draw rate recommendations were as follows:

- Convergence rate < 0.8 mm/day – no change in draw quantity.
- Convergence rate 0.8–2.0 mm/day – increase draw rate by 22–33 tonnes/shift up to maximum 110 tonnes/shift.
- Convergence rate > 2.0 mm/day – increase draw rate by 44–66 tonnes/shift up to maximum of 190 tonnes/shift.

If the convergence rate was 1.5–3 mm/day for more than 3–7 days even after increasing the rate from a specific drawpoint, the draw rates were increased from the neighbouring drawpoints. The effect of increased draw rate on convergence was monitored daily and as soon as the convergence rate slowed down, the rate was reduced back to the standards rate.

11.2   BEHAVIOUR OF FRAGMENTED ORE – A CASE STUDY FROM A SUBLEVEL CAVING MINE

The Perseverance Mine was located 370 km north of Kalgoorlie and about 800 km northeast of Perth. The orebody was discovered in the 1970s, and in 1989 an open pit was established. The Perseverance pit was completed in 1995 to a depth of 190 m. Since 1995, the orebody was mined exclusively by sublevel caving (SLC) method. In 13 years of operations, the mining depth at the Perseverance Mine has progressed from 375 m to 900 m below the surface (i.e. about 40 m per year). Production levels were 25 m apart (floor to floor) with ore crosscuts being 5 m by 5 m. Typically, the orebody is 80 m wide and 150 m long. The dip of the orebody is sub-vertical with an inflection from 420 m to 620 m below surface, where the dip flattens to 45° before steepening again below. The hangingwall contact with ultramafic orebody is marked by a very prominent shear zone containing low shear strength metamorphic minerals. The mineralogical, lithological and structural complexity of the Perseverance deposit gives rise to a variety of different rock mass domains and behavioural responses (Szwedzicki et al., 2007).

11.2.1   Ore fragmentation and size distribution

A classification system was developed to provide a population size to suit operational requirements. The classification included fines (< 0.03 m), medium (0.03–0.1 m), coarse (0.1–0.3 m), large (0.3–0.9 m) and oversized fragments (> 0.9 m). Fines represent material that, at certain values of moisture, can develop cohesive forces and thus form hang-ups in orepasses and drawpoints. The large fragments represent material that can be handled by LHDs, while an "oversize" category (blocks that are larger than 0.9 m) represents the material that typically requires secondary blasting.

A cumulative distribution curve from average fragment sizes is shown in Figure 11.7. The analysis of the size distribution indicated that the blasted ore is very well fragmented with, on average, 30% of fragments of less than 0.05 m, 50% of less than 0.1 m and 75% of less than 0.3 m. The average number of oversized blocks (measuring > 0.9 m) was 0.7 blocks per drawpoint ring (Szwedzicki and Cooper, 2007).

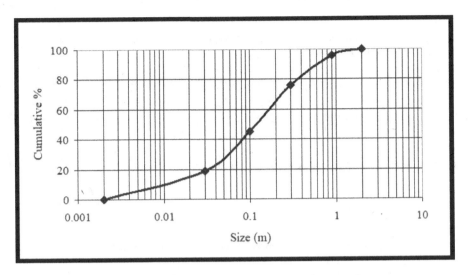

Figure 11.7 Cumulative distribution of fragment sizes – average values from two levels (Szwedzicki and Cooper, 2007).

11.2.2   Ore recovery

A draw marker study aimed at improving metal recovery by gaining an understanding of the ore flow characteristics in the cave, determining the dimensions of the extraction zone and ascertaining the extent of the secondary and subsequent draws. The markers were fabricated from steel pipe, 250 mm long and 45 mm in diameter. Each marker had an individual welded identifier and was filled with concrete to ensure that it had a similar density to that of the ore. Markers were inserted and grouted in holes drilled in rings from the crosscuts (2 to 11 holes used per ring). The trial was conducted over 12 rings, in five crosscuts, on three different levels (Hollins and Tucker, 2004).

Over the four years of the trial, extraction of the SLC progressed down through four levels (i.e. 100 m). Out of 1,760 markers installed, 741 of them were recovered, with the recovery of the markers exhibiting substantial variability. The identified causes of these variations included single and multiple rings being blasted at the same time, hang-ups due to large blocks entering drawpoints, piping of ore flow during the initial draw and varying amounts of ore drawn from different drawpoints. It is also believed that some installed markers passed through undetected.

On average for the five trial locations, it is estimated that in areas where only one ring was blasted and drawn at the time, 60–70% of the recovered markers were found on the level on which they were installed (primary recovery), 20–25% were recovered on the subsequent level (secondary recovery), 10–15% were recovered on the third level (tertiary recovery) and up to 8% were recovered on the fourth level (quaternary recovery).

11.2.3   Primary recovery

Primary recovery takes place on the level on which the production ring is blasted (i.e. the same level on which the markers were installed). The average assessed width of the draw zone was only 10–11 m. With a crosscut width of 5 m and a pillar width of 9 m, the width of the draw zone had to be at least 14 m to achieve interaction between the draw zones. No single marker was recovered in a neighbouring drawpoint on the same level. The trial, therefore, did not confirm interaction between draw zones on the same level.

It was concluded that draw from the drawpoints often developed as narrow "pipes", which extend to the next level within the drawing of only 15% of the design tonnes. When this occurred, waste often mixed with primary ore and reported to the drawpoint. It was also reported that hang-ups often formed at about 30% of the draw tonnage. These hang-ups were thought to be caused by large blocks of rocks reporting from the levels above, or large blocks formed due to deficiencies in blasting at the periphery of the rings.

The depth of draw on the level usually reached, but did not exceed, 3 m. When two rings were fired together, no material was recovered on that level from the ring furthest from the brow.

11.2.4 Secondary recovery

Ore drawn on the level immediately below the blasting level is classified as secondary recovery. The secondary recovery typically flows from the "pillar" between the drawpoints on the level above, which was typically 3–4 m wide (in one case it was 7 m wide). It was observed that secondary flow was frequently "disturbed", as indicated by the mixing of material (markers) from different heights and depths (i.e. the markers reported to the drawpoints in a mixed order). An example of the interpretation of zones of primary and secondary draw is shown in Figure 11.8. Contours of recovered markers delineate primary recovery. The markers outside the contours are considered to be in the zone of the secondary, tertiary or even quaternary recovery.

The secondary recovery markers typically started reporting to the drawpoints on the level below at about 20% of the designed draw tonnes. This confirmed the previous estimations that the dilution entry point was at 20% draw of the designed tonnes.

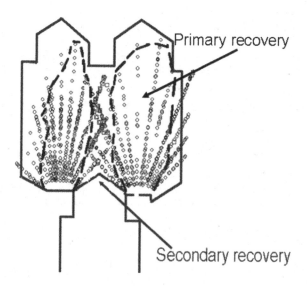

Figure 11.8 Contours of primary and secondary draw zones (Szwedzicki and Cooper, 2007).

11.2.5 Tertiary and quaternary recoveries

Markers recovered two levels below that from which they were blasted are classified as tertiary recovery, and those recovered from three levels below as quaternary recovery markers.

In areas where only one ring was fired and subsequently drawn, tertiary recovery was between 10% and 15%. When two rings were fired together (e.g. due to unfavourable drilling or charging conditions), ore from the ring furthest from the brow was commonly recovered during the tertiary draw. As a result, most of the tertiary markers were recovered from areas where multiple ring firings occurred. In areas where two rings were fired simultaneously, tertiary recovery increased to 30%.

Overall, quaternary recovery was up to 8% of the material drawn from the drawpoints, while certain areas exhibited no recorded quaternary recovery at all. However, observations indicated that material can flow to lower levels over substantial distances. For example, a steel set has recovered on a level 270 m below the level in which it was installed.

11.2.6 Effect of rock mass properties on fragmentation

It was noted that the size distribution of the fragmented rock mass was related to the mechanical properties of the rock mass. At Perseverance Mine, three types of mineralisation could be distinguished in the ore zone. Their mechanical properties, mining rock mass ratings (MRMR) and a relationship between the rock mass properties and fragmentation size was established (Table 11.1).

11.2.7 Reduction of fragment size with tonnage drawn

As the ore flows through a draw column, rock blocks undergo additional fragmentation due to stress, abrasion and attrition, and the fragment size is thus reduced. As the blocks move down, the abrasion against other blocks also results in the rounding of corners, with the chipped small particles from these rocks forming fines. Typically, angular fragments are associated with blocks originating from the collar of the draw column, while rounded fragments are expected to have travelled a larger distance before reaching the drawpoint.

Examples of rock fragments that were collected from primary draw and from secondary draw on the subsequent level are shown in Figures 11.9 and 11.10 respectively. An example of a relationship between the cumulative size distribution of fragments and the amount of material drawn is given in Figure 11.11.

To determine the reduction in size as ore was drawn through the SLC, statistical analysis was carried out for all drawpoints. The percentage of fine fragments was found to change with the tonnage extracted. As the tonnage drawn increased, the percentage of fines increased

Table 11.1 Rock properties and rock mass classification

Rock type	Uniaxial Compressive strength (UCS (MPa)	Uniaxial Tensile Strength (UTS) (MPa)	MRMR	Average fragmentation, m
Talc-chloride	46	4.3	< 25	0.05
Serpentine-talc	90	7.1	40–60	0.1
Olivine-serpentine	126	12.1	> 60	0.2

Figure 11.9 Coarse rock fragments recovered during the primary draw (Szwedzicki and Cooper, 2007).

Figure 11.10 Rounded rock fragments that travelled two levels (Szwedzicki and Cooper, 2007).

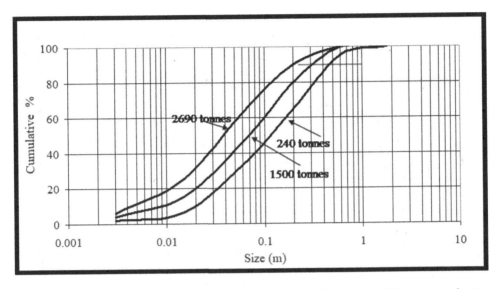

Figure 11.11 A relation between a cumulative size distribution of fragments and the amount of material drawn (Szwedzicki and Cooper, 2007).

(Figure 11.11). Within the first 1000 tonnes, the percentage of fine fragments (< 0.03 m) typically varied between 5% and 20% (with an average of 15%). Between 1000 and 5000 tonnes, the percentage of fines present varied between 5% and 35% (with an average of 20%). Between 2000 and 5000 tonnes drawn, all drawpoints had percentages of fines between 10% and 40% (with an average of 25%).

The percentage of medium and coarse fragments, within the size ranges of 0.03–0.1 m and 0.1–0.3 m, remained relatively constant throughout the drawing process, with values from 10% to 40% in each size range. However, the percentage of large fragments, within the 0.3–0.9 m size category, exhibited a wide range, with observed values between 5% and 60%. The majority of the drawpoints had between 10% and 30% of fragments within this size category.

Drawpoints that had exceeded their designed ring tonnage (typically more than 3000 tonnes drawn) had on average only 0.3 oversized fragments. The percentage of oversized fragments varied between 0% and 15% for these drawpoints, with the majority of drawpoints having less than 5% oversized fragments. Secondary blasting to reduce the oversized fragments was required for about every 1000 tonnes of ore.

The change in fragmentation, as draw from the SLC progressed, was also monitored at specific drawpoints. Figure 11.12 shows an example of the reduction in fragmentation. The MRMR in the area was between 40 and 60. Observations were made after drawing 240, 1490, 2090 and 2690 tonnes. Between 2090 and 2690 tonnes drawn, the size distribution remained relatively constant. The reduction in fragmentation size was almost linear for the first 2000 tonnes, with the effects of secondary fragmentation diminishing with the tonnage drawn. It is interesting to note that for the first 300 tonnes, approximately 50% of fragments were smaller than 0.1 m, whilst at the end of draw (at 2960 tonnes), almost 85% of fragments were smaller than 0.1 m. Fragmentation was analysed to determine its impact on recovery. It was concluded that fragmentation has little to no influence on recovery.

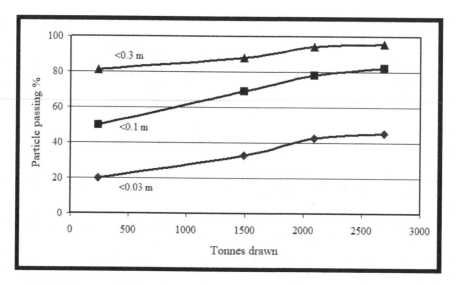

Figure 11.12 Reduction in the fragment size with the tonnes drawn (Szwedzicki and Cooper, 2007).

A four-year marker trial was conducted at the Perseverance Mine to characterise ore flow in sublevel caving conditions. The movement through the SLC of over 700 markers was analysed and supplemented by a fragmentation study. From this analysis it was concluded that:

- For very well-fragmented ore, the width of an extraction zone was 5–6 m wider than the width of the drawpoint. With the pillars between crosscuts designed to be 9 m wide, no draw interaction took place between the neighbouring extraction zones.
- When a single blast hole ring was fired and subsequently drawn down, primary recovery was 60–70%, secondary recovery was 20–25%, tertiary recovery was 10–15% and quaternary recovery was up to 8%.
- When two rings are fired together, during primary recovery no material was recovered from the ring furthest from the brow; however, tertiary recovery increased to 30% from subsequent extraction levels.

11.2.8   Draw management practices

The objective of effective management of the ore draw is to maximise draw before dilution enters drawpoints, delay dilution entry as long as possible, and correct anything that contributes to early dilution entry (Bull and Page, 2000). These objectives have been achieved over the years by developing practices and guidelines that were formulated in the cave management practices referring to draw control, which are given below:

Do not draw below shut-off grade

The quantity and value of the dilution material affects the economics of the operation. When a ring is blasted, the first part is largely drawn clean (no dilution). Dilution from above and behind the ring then starts to come into the drawpoint. This mixture of ore and waste from levels above determines the grade of the material being drawn. The proportion of waste increases as the tonnage drawn from the drawpoint increases, decreasing the drawn grade, and continues until the shut-off point is reached. To prevent any mixing of waste with ore as it is drawn to the extraction points, the grade could be reduced to 90% of the *in situ* grade. With uniform draw, it allows to keep an ore blanket that should be maintained. It results in improved extraction grade.

Extraction should attempt to uniformly draw down this ore/waste interface entry into the drawpoints. This draw is best controlled by interactive drawing from neighbouring drawpoints.

Do not draw too fast or for too long

Drawing a drawpoint too fast causes the formation of a tall narrow draw zone (a pipe) leading to surging of fines dilution from above or behind the broken ring, resulting in the lowering of the draw grade and reducing the tonnage that can be economically drawn. Once piping has occurred, ore from the sides of a ring cannot be drawn down easily, and dilution from higher up in the pipe is drawn down instead. The gradual, or incremental, drawing of a drawpoint reduces the risk of piping and allows ore from the sides of a ring to slowly enter the draw zone. The practice was to restrict draw for the first two shifts following a ring firing to 400 tonnes per shift, and then increasing draw rates to 800 tonnes per shift thereafter. Drawing for too long from a drawpoint ring, or overdrawing, can negatively impact overall production rates from the SLC by extending the life of the level, in turn affecting the sequence of the crosscuts. Additionally, extending the life of a drawpoint may lead to brow wear and blast holes in subsequent production rings becoming dislocated or closed off.

Do not continue drawing when flow is obstructed

Obstructions in a drawpoint may include large rocks, sheets of mesh and shotcrete, or other ground support that has not fired off completely. It may also include material that has been drawn into the cave zone from the levels above. In the case of large rocks, obstructions can be present either at the brow (low hang-up) or further up inside the drawpoint (high hang-up). When there is an obstruction, the uniform flow of material is interrupted, and preferential draw of fine material will occur. It may also mean that only one side of the drawpoint is being drawn, providing a similar effect. This could lead to piping and early dilution entry, with waste ingress then causing rings to be shut off earlier than planned. Clearing obstructions from drawpoints immediately is critical so that normal flow is not interrupted, and piping does not occur.

Draw across the full drawpoint width

Narrow draw widths result in earlier dilution entry. Drawing only from the centre, or from only one side of a drawpoint, results in the effective draw column width being reduced. For a more even draw and less dilution, it is important to draw across the full width of a drawpoint, alternating left and right for each full bucket. Drawing across the full drawpoint width encourages ore to flow evenly, which helps to minimise early dilution entry and maintains an even boundary of the ore/waste interface.

Pull from drawpoints to loosen material immediately before charging

During the caving process, material in the cave consolidates with time if left undisturbed. This means that each blast will take place against tightly compacted cave material which may lead to the formation of "frozen ground". Drawing material from the previous ring just prior to blasting helps to loosen up the cave material and creates the required void necessary for a successful blast. Experience has shown that only 100 tonnes are required to be drawn to loosen the cave material sufficiently.

11.3   ANALYSIS OF A DIAMETER OF A DRAW ZONE FROM TWO CASE STUDIES OF CAVING MINES

Diameter of the ore flow zone was determined in two case studies. One was in a sublevel caving (SLC) mine, where markers were installed in the rock mass to determine the ore flow, and the second was in a block cave (BC) mine, where many tramp materials from the block above was recovered from the drawpoints on the extraction level of the lower block. To calculate a diameter of the draw zone, it was assumed that the draw zone was cylindrical with the height being determined by initial position of markers/ramp material. The diameter was calculated by determining the volume of the material drawn before the markers/tramp material reported to the drawpoint below. The relationship between the tonnage drawn and volume was calculated using the *in situ* density of the various rocks and the swell factor of the caved rock.

In a sublevel caving mine, with a crosscut width of 5 m and a pillar width of 9 m, the width of the draw zone had to be at least 14 m to achieve interaction between the draw zones. The average assessed width of the draw zone was only 10.5 m.

In a block caving mine, with the drawpoints on 18 m centres along the panel, the diameter of the draw zone was 6.3–14.6 m, with an average diameter being about 12 m. In one case, the calculated diameter of the zone was only 6.3 m, which indicated that void diffusion mechanism took place rather than mass flow (Brown, 2007).

It was noticed that irrespective of various geological origins of ore, i.e. various ground conditions, the diameter of the draw zones was found to be of similar range (Table 11.2).

Table 11.2 Draw parameters

Parameters	Sublevel caving	Block caving
Height of the considered zone, m	10-25	320
Tonnage of material drawn from every drawpoint, t	< 5,000	20,000–150,000
Drawpoint width, m	5	3.6
Diameter of the draw zone:		
Min, m	7.2	6.2
Max, m	13.0	14.6
Average, m	10.5	12
Difference between diameter of draw zone and drawpoint width, m	5.5	8.4

Despite various draw geometry, different draw times and different fragmentation, results obtained indicated that the average width of the draw zone was similar – 13 m for SLC and 14.6 m for BC. It was not observed that interactive draw took place. However, in both draw conditions, the minimum calculated diameter of the zone was only 6.3 m (for BC) and 7.2 m (for SLC), which indicated that void diffusion mechanism also took place (rather than mass flow).

11.4 BEHAVIOUR OF FRAGMENTED ORE IN OREPASSES

Orepasses are used to move broken ore or waste rock from a higher level of a mine to a lower one. They may also be used to store broken material. The proper functioning of orepasses is critical to assuring ore flow, and hence production in underground mines. Operation of ore-passes must ensure that broken rock flows when required. If the passes become inoperable because of hang-ups or blockages, the production of the mine can come to a halt. Hang-ups can present safety problems in their removal and can significantly reduce orepass life.

Hang-ups occur commonly in passes; for example, information from 3600 passes in gold mines in South Africa shows that 35% of the passes were subject to hang-ups or blockages (Emmerich, 1992). Stability of the hang-ups is uncertain, as they can suddenly crash down. The clearing of such hang-ups so that the flow can be re-established is often very hazardous. Clearing of hang-ups might be accompanied by runaways, i.e. uncontrolled flow of the material past the control chute, and that includes mud rushes.

According to statistics published by the South African Department of Minerals and Energy over the period of 1999–2004, there were six reportable rock pass accidents and three deaths per year related to orepasses (Stacey and Erasmus, 2005). Efforts to restore flow of ore are often hindered by lack of suitable clearing methods. Hang-up removal is often based on the experience of operators, i.e. what has worked in the past. Because of different rock types, geological structures and orepass design, hang-up removal methods that worked in one mine may not work in another or may even not work in different areas of the same mine (Stewart *et al.*, 1999). Despite substantial advances in mechanisation and automation of many min-ing activities, hang-ups still occur frequently in underground mines and methods for their removal have not been widely researched. This review deals with the type of hang-ups that occur most frequently in orepasses. The review is intended to cover scenarios that typically take place in underground mines; however, it must be recognised that each hang-up forms under different conditions. Ultimately, it is a competent person that must undertake a formal risk assessment and decide on the appropriate method to use. The purpose of this review is to ensure that if a pass hang-up occurs, the remedial work is carried out using safe and appropriate methods.

11.4.1 Types of hang-ups

A distinction is made between blockages at the bottom of the orepass and hang-ups that are found above the brow. Impediments to flow in the feeding zone or along the length of an ore-pass or drawpoint are defined as hang-ups, while blockages are localised at the bottom zone (Hadjigeorgiou *et al.*, 2005; Lessard and Hadjigeorgiou, 2003a). Flow of material in ore-passes depends on the properties of the material, i.e. size distribution, amount of fines, cohe-siveness of fines, density, angle of friction and water content. The ore and broken waste flow process is driven by gravity and resisted by friction and cohesion, so consequently hang-ups form when friction or cohesion forces reach critical values. When friction increases, interlocking hang-ups form; when cohesion increases, cohesive hang-ups occur.

Hang-ups due to interlocking rock fragments

Hang-ups occur when larger rock fragments form stable arrangements like arches or domes in the orepass. Arching/interlocking occurs when large rocks wedge themselves. Formation of arches depends not only on the maximum particle size but also on the particle size distribution. They frequently occur at constrictions (e.g. reduction in diameter below enlargement of the pass or at chutes); however, they can form anywhere along the orepass. For the dry feed, the most probable hang-ups are of the interlocking type.

Hang-ups due to cohesive arches

Cohesive arches form due to fine particles adhering to each other, as fine particles exhibit a cohesive component to their shear strength in addition to a frictional component. Fine material, due to cohesion, loses its ability to move. This resistance is enhanced if moisture is present. When there is no movement of the material in the orepass, the possibility of cohesive arches forming increases. Cohesive arches tend to form if material is not moving for some time.

Hang-ups due to interlocking of blocks and cohesion

It often happens that interlocked arches are later additionally cemented due to cohesion. Even if a small amount of water is seeping through the orepass, and large blocks wedge forming an arch, small fragments and fines can be transported and accumulate around constrictions caused by arched blocks. Later the fines may cement and develop cohesion, which adds force to the existing frictional forces. In the case of a mixture of large rocks and fines that are dumped into an orepass, large rocks produce compaction of fines and formation of cohesive hang-ups. These types of hang-ups are the most difficult to clear.

11.4.2 Formation of hang-ups

Gravity flow of fragmented rocks is influenced by material properties and the size and shape of the orepasses. Hang-ups/blockages can and do occur when the size of the material is incompatible with the size and configuration of the passage or draw practices are not well established and/or adhered to. In large orepasses due to impact, wear and deterioration in ground conditions, it happens that large rock blocks detach from walls, blocking the bottom part of the pass.

It appears that very little information on the physical properties of the ore or waste rocks exists regarding size distribution, amount of fines, density, angle of internal friction and how they relate to material flow or how they contribute to the formation of the hang-ups. If the ore contains significant quantities of fines, the active flow channel in the orepass could reduce in diameter, forming so-called rat holes.

From a geometry point of view, the hang-ups are more likely to take place when material mass flow changes to funnel flow at a constricted area. Factors predominantly contributing to formation of hang-ups are cohesion, consolidation and compaction of fines and constrictions in the orepasses. The formation of hang-ups depends on the percentage of fines and oversized fragments, the shape of the fragments, the relative size of large blocks to the size of the orepass, and the velocity of the flowing ore and compaction on free fall.

Block size

To ensure material flow in an orepass by preventing interlocking hang-ups, a minimum ratio of the orepass dimension over the particle size dimension should be maintained. Laboratory testing and field observations show that occurrences of interlocking hang-ups are dependent on the size distribution as well as on the absolute size of the material.

Review of publications indicates that the recommended ratio of maximum rock fragment size to orepass diameter varies within a wide range, from 1:3 to 1:6. A ratio of the orepass diameter to maximum particle size of 5 is likely to result in flow, whereas a ratio < 3 is likely to result in a hang-up. A ratio between 3 and 5 is likely to result in flow (Beus et al., 2001; Hambley, 1987). Numerical modelling (Lessard and Hadjigeorgiou, 2003b) indicates that the ratio should be > 2.8. From an ore flow point of view, oversized rocks can be defined as blocks with a diameter 0.20% of the diameter of the pass. Grizzlies are important in preventing block type hang-ups because they control maximum rock fragment size in the material dumped into the orepasses.

Cohesion

The fines, due to cohesive forces, worsen the flow characteristics of broken rocks. Cohesion of fine material can hold the larger particles together and form a continuous arch or dome across the orepass. The finer the material, the greater the potential for cohesive arches. The fines can form a continuous matrix in which the coarser particles are embedded. The cohesive arches may be formed when fines constitute a part of the feed and when they are moist.

However, due to different flow characteristics, fines can move faster than larger fragments, and segregate and accumulate in certain positions. It may happen that the last few buckets of ore taken from a stope have a higher percentage of fines. As a result, even a smaller amount of fines can result in cohesive arches. Soft rocks travelling through long orepasses can undergo comminution, resulting in the generation of a substantial amount of fines by attrition (Goodwill et al., 1999). As a result, feed material that is relatively fines-free, when tipped may form cohesive arches at the bottom of a long orepass due to a large amount of fines. The value of cohesion depends on size and fragmentation of the fines and on the water content. Dry fines or fully saturated fines do not have any cohesive forces. However, even a small amount of water introduced by spraying for dust control can substantially mobilise cohesive forces.

Consolidation

Certain sulphide minerals, when in contact with air and water, in time, can consolidate and cement. Prolonged time between draws allows for such consolidation. Keeping the orepass active by frequent draw downs prevents such phenomena. The maximum period without draw down depends on the susceptibility of minerals to consolidate, in combination with temperature and moisture, so it should be determined for each site. Additionally, consolidation of the fine material occurs when it dries out in the orepass. Consequently, there is usually an urgent need to clear blockages as soon as they are identified.

Compaction

Compaction of the material can take place due to the impact of the fall of material from a tipping point or due to the load exerted by a high column of fragmented rocks. Fragmented material, when compacted, develops high frictional forces that resist the free flow of ore or waste rocks. A height of material more than twice the diameter of the orepass provides a cushioning effect, reducing compaction of material at the bottom of the orepass.

Constrictions

Constrictions or reductions in a cross-section of the pass result in the restricted flow of material and a change in flow mechanism from mass flow to funnel flow. Constrictions can be caused by a change in the shape of a pass, blockages by slabbing of walls or by blockages caused by foreign material. Constrictions are likely to contribute to the recurrence of hang-ups. Constrictions resulting from change in the shape can be due to the rough wall of the orepass, orepass enlargement (due to blasting or scaling) or structural features at the bottom of the orepass, like chutes. Constrictions can also be formed by cohesive material sticking to the wall of the pass causing the formation of "rat holes". Blockage by slabbing may occur when large slabs of rock slough and detach from walls of orepasses. Such structural or stress failures result in an orepass enlargement and may lead to an orepass collapse.

Keeping the orepasses full provides a degree of confinement that contributes to its stability. Blockage by foreign material in large-scale mechanised mining may occur when large-sized foreign material such as steel support, rock bolts, wire mesh, etc. can enter the pass. That material can block material flow, especially at constriction points. Though hang-ups may occur due to random causes, e.g. blockage by foreign material, mining practice proves that the majority of hang-ups are from repetitive causes which may result from water seepage into the pass, feeding fines from the same part of the orebody or constriction caused by a change in the shape of an orepass. It has to be noted that in most cases a number of superimposed causes leads to the formation of hang-ups or blockages.

11.4.3   Hang-up prevention

In properly designed and maintained orepasses, there are two main reasons for the formation of hang-ups:

- large blocks (oversized rocks) fed into an orepass at the same time can result in interlocking and wedged blocks, and
- wet fines ("sticky mud") can form compacted and cemented arches when the draw from an orepass is discontinued (even for a single shift).

To prevent the formation of hang-ups, the following operational practices should be applied:

- Continuous drawing of material from pass – Material within the orepass that contains a high content of fines should be constantly moving. An interval between the consecutive draws should be as short as practicably possible. When the potential for disruption of drawing exists, ore should not be dropped down the pass.

- Breaking of oversized blocks on grizzlies – A large number of oversized rocks should not be passed through a grizzly at the same time (e.g. end of a shift). Despite breaking by a rock breaker, several large blocks tend to form a hang-up. Oversized rocks should be gradually sent through a grizzly with no more than three blocks sent down the grizzly in between buckets of well-fragmented rock.
- Reduction of excessive water – Water when mixed with fines forms "sticky mud" that blocks the orepasses. It is important that water used for washing walls after blasting and for dust suppression is used prudently. Excessive water has a detrimental effect on fines. This applies to all drawpoints and to the dust suppression system.
- Keeping low level of ore in orepass – If there is a large amount of fines, the level of the orepass should be kept low so that if there is a hang-up, it is near the bottom where it can be easily removed, for example, by high-pressure water jetting.
- Blending of wet and fine material – If practical, material that is fine and wet should be blended with coarser rock, for example, one bucket of wet fine material and one bucket of coarse material.

Chapter 12

Mitigation of rock mass response through geotechnical quality assurance

Rock mass response to mining activities can be affected by the quality of development work and production. A quality assurance program in ground control management is achieved by developing and implementing ground control policies and management system. The system is effective when authority and responsibilities are clearly specified and delegated to (a) competent person(s).

The system should be documented and must specify agreed actions and a method of approval and verified of ground control activities. All ground control activities such as data collection, drilling, blasting, maintenance of excavations and monitoring should be described in the work procedures. Geotechnical inspection and monitoring should allow for preventative and corrective action to be taken before rock mass becomes unstable.

Safety and cost efficiency in open pit and underground mining operations are directly related to geotechnical activities. The activities must follow recent research developments and must cover geotechnical management systems, document control, communication, geotechnical input into mine design, support installation or geotechnical risk management (Szwedzicki, 1989a).

Ground instability often results from deficiency in quality in ground control activities. For example, ground control procedures might become inadequate to manage risk over time, and improper practices or faulty materials and equipment might be accepted. Consequent instability of the rock mass represents a considerable safety-related problem and may result in economic losses.

Safety and cost efficiency of mining operations could be substantially increased by implementing a policy on quality assurance in ground control management. Increased safety in mines results from improved rock mass stability; improved geotechnical design process; efficient monitoring, inspection and reporting; and an increase in the quality of support. Reduced production cost can be achieved by eliminating blast damage to the rock mass, better rock fragmentation, reduction in the number of oversized rocks and reduction of support requirements.

The main objective of a quality system in ground control management is to produce information and then use that information for improvement in mine safety (through increasing the stability of the rock mass) and reduction of production (mining and milling) costs (Szwedzicki, 2003a).

The ISO 9000 series (ISO, 9001; ISO, 9002; ISO, 9003) provides guidance in developing an effective quality system that can be integrated into a geotechnical management system.

12.1   QUALITY IN GROUND CONTROL

Ground control is defined as the ability to predict and influence the behaviour of rock mass in the vicinity of underground or open pit mining excavations. It consists of recognising ground conditions, geotechnical mine design, mining activities (drilling, blasting and support), and monitoring rock mass performance. Quality represents features and characteristics of the activities that bear upon its ability to satisfy stated or implied needs. In simple terms, in relation to ground control, quality is conformance to requirements or specifications provided in the following:

*   legislation
*   company documents such as procedures, rules, standards, or codes of practice
*   manufacturers' and suppliers' instructions
*   approved best work practices

Quality control is the operational techniques and activities to fulfil requirements for quality, while quality assurance is defined as planned and systematic actions to provide adequate confidence that activities will satisfy requirements for quality.

Stages of implementation of the quality program in ground control include (after ISO, 9002):

1   Preparation of a policy on quality assurance in ground control

The quality policy refers to management commitment and is a description of company objectives. The policy objectives should be quantifiable and measurable. The policy should be supported by procedures that are documented methods of carrying out ground control tasks. Policy and procedures should be relevant to the whole organisation and should be approved by an authorised person.

2   Appointment of a management representative.

A competent person(s) should be appointed and should be given authority and responsibility for all ground control activities.

3   Development of a specific document, for example, a Ground Control Management Plan.

The Ground Control Management Plan introduces quality measures in each step of the mining activities and should focus on key control points, i.e. data collection, design, mine production, and instrumentation and monitoring, and geotechnical risks management. The document should specify responsibilities and geotechnical actions with time frames.

4   Determination of objectives and targets.

Work procedures are standard documents showing each step of a job or a task, the hazards identified with each step and actions to be adhered to in order to manage and control each hazard. Specific quality parameters and acceptance criteria should be included in all work procedures. The criteria should specify acceptable deviation from the standards.

5   Implementation and control of all activities.

For the system to be operational, all ground control activities must be implemented and controlled by periodical reviews or audits. Actions arising should be verified in the agreed timeframe.

6   Review.

The quality program should allow for feedback and be periodically reviewed.

The ground control quality system at a mine should comprise geotechnical management (organisational structure, authorities and responsibilities), documentation (procedures, practices, instructions and specifications), operational activities (drilling, blasting and support installation) and monitoring and inspection. Ground control quality assurance should cover present activities and future requirements.

12.2   QUALITY ASSESSMENT

Quality can be described by a quality grade and a quality level (Fox, 1995). The grade is specified by standards or specifications, for example, rock bolt protrusion of 0.1–0.3 m (low grade) or 0.1–0.15 m (high grade). The level can be specified by a number of bolts installed according to the specification, for example, 99 out of 100 (high level) or 90 out of 100 (low level).

Quality grade is defined in quality policy, procedures or specifications, while quality level is achieved in an implementation phase, i.e. during geotechnical activities.

Measuring ground control quality can be achieved by recognising a shortfall in the following:

- quality grade, i.e. discrepancy between company specifications (requirements or practices) and standards, best work practices or manufacturers specifications
- quality level, i.e. discrepancy between the requirements set in geotechnical procedures or specification and actual implantation of the activities

Such discrepancies can be recorded and counted as follows:

- number of deficiencies recognised (quality grade and level)
- number of items found deficient (quality level)
- measured deviation from the standard (quality level)

Assessments of ground control quality assurance can be achieved through audits and management reviews. An audit is a systematic and independent examination to determine whether activities and related results comply with planned arrangements and whether these arrangements are suitable to achieve objectives. The objective of an audit on quality assurance in ground control is to provide mine management with the information on status and potential improvement in geotechnical activities. The audit also covers safety and risk management aspects of mining operations related to the ground control management.

There are two types of audits – a system audit and a compliance audit. The system audit is used to determine the existence and validity of the ground control management system. The compliance audit is used to confirm whether specified procedural practices in geotechnical planning and design, ground control activities and inspection and monitoring are actually being implemented and are effective.

A system audit on quality assurance in a ground control management system should seek evidence of the following:

- clearly defined responsibilities and authorities
- documented procedures, practices and instructions
- knowledge and understanding of responsibilities, authorities, procedures, instructions, etc.

A compliance audit on quality assurance in ground control management system should seek evidence of the following:

- correct operational procedures approved by the authorised person
- adequacy of personnel, equipment facilities and general resources
- effectiveness of the system when correctly operated

The audit cannot examine all activities but chooses random samples and examines them for non-compliance or for possible improvement. The audit is not an appraisal activity or process, but an action taken to prevent the recurrence of any deficiencies discovered. When the evidence collected indicates that the requirements of the procedures and standards are not being followed, this should be recorded as non-conformance. Management reviews and audits can highlight areas for potential improvement. Action plans are then developed to address identified issues.

A program of the audit on quality assurance in ground control management could include the following techniques:

- interviews and discussions with line managers responsible for ground control activities
- inspections of work places in the open pits or underground mines
- review of ground control documents, standards, work procedures and practices
- discussion on and review of geotechnical input into mine design
- observation and monitoring of quality of drilling and blasting
- observation and recording of rock mass behaviour and modes of failure
- interviews and discussions with supervisors and operators responsible for drilling, blasting and barring down
- review of geotechnical records and data
- discussions with geotechnical, mining production and other technical staff
- development of action plans to address identified issues

The audits can highlight the present achievements that meet international mining standards but also may disclose deficiencies and may reveal directions for further improvement.

12.3   QUALITY ASSURANCE IN GROUND CONTROL MANAGEMENT SYSTEM

The ground control management system should ensure that an organisational structure exists to manage and verify quality in ground control. A management system can be evidenced by the presence following system components:

12.3.1   Responsibilities and authority

A competent and suitably qualified person(s) must be appointed to manage, supervise and perform ground control activities such as assessment of ground stability, geotechnical input into drilling and blasting, maintenance of excavations, monitoring and inspections. Responsibilities and authority must be well defined and should be reflected in respective job descriptions.

12.3.2   Compliance with legislation

A management system must be created so that all ground control activities are carried out in compliance with mining safety legislation. The system has to ensure that provisions of the Act and Regulations are followed.

12.3.3   Competency and training

Competency and training system must be developed to ensure that all employees responsible for ground control activities are competent (i.e. trained, qualified and experienced) in reading ground conditions, detecting signs of ground instability and carrying out barring down of the exposed ground. The system must ensure that drillers, blasting crews and support installation operators are competent in their duties. It must also ensure that supervisors and professionals are continuously trained and exposed to newly introduced procedures and practices. Relevant records of education and training should exist for all employees involved in ground control.

12.3.4   Communication and reporting

A communication and reporting system between various levels of a management structure and between all professionals involved in ground control issues should be established and enforced. Communication can be formal (e.g. written instructions, memoranda) or informal (e.g. verbal during meetings). The communication system should ensure that all interested parties receive the required information and that information is understood. Proper communication channels should allow for effective feedback.

12.3.5   Document control

A document control system requires that all ground control policies and procedures are approved, distributed, reviewed and archived. Document distribution and circulation must follow an approved list to ensure that all relevant personnel are advised and have been provided with access to relevant documents. The system must prevent documents from being withheld or put aside.

A suitable person should be responsible for revisions and must implement a system that allows for timely withdrawal of old documents and replaces them with the latest versions. All written procedures and practices should have a revision date by when they must be discussed and reviewed. If needed they should be updated or modified. All changes must be readily available and be communicated to personnel that might be affected.

A record keeping, and archiving system has to be developed to prevent misplacement or loss of documents, consultants' reports, collected data, etc. Records should be readily accessible when required.

12.4   Quality assurance in geotechnical planning and design

Geotechnical design criteria should be established and a system of geotechnical input into mine planning and design should be implemented in the early life of the mine. Results of the investigations into ground conditions should form a base for selection of the mining methods, support selection and design of geotechnical instrumentation.

12.4.1　Data collection and analysis

A system should be in place to ensure that all needed geotechnical information and data are systematically collected, processed, interpreted, analysed, documented and archived. The information and data should be collected according to accepted standards or well-established methods. Changes in ground conditions or behaviour have to be monitored and reported. Figure 12.1 gives an example of collapse of an access ramp along a structural fault that was not detected.

The collected data and information must serve a purpose and must be analysed. The results should be used for design, planning, support installation and the installation of geotechnical monitoring.

12.4.2　Geotechnical planning

A systematic approach to mine planning and design should be based on geotechnical engineering methods. Geotechnical planning should take into account the life of each excavation and life of the mine.

12.4.3　Ground Control Management Plan

A Ground Control Management Plan should be the leading geotechnical document that describes the ground control system and includes all relevant geotechnical information. The document should include important geotechnical data, specify minimum standards of ground

Figure 12.1 A failure of a ramp in an open pit along unidentified structural feature (Szwedzicki, 2003a).

control, refer to procedures and consultants' reports and give an overview of geotechnical settings, for example, classification of rock mass and delineation of geotechnical domains. It also should include identification and evaluation of geotechnical risk. The document should include a schedule of monitoring and short- and long-term plans of geotechnical activities. The plan has to be approved by management and reviewed annually or more frequently if necessary.

12.4.4 Mine closure

A Ground Control Management Plan should cover the whole-of-mine life, including a mine closure phase. Provision for geotechnical aspects of mine closure should be addressed, for example, final wall angle, long-term stability, securing access to the site and possible surface subsidence or erosion (Szwedzicki, 2001b). The provision should also include geotechnical monitoring and observations after mine closure.

12.4.5 Geotechnical design

A responsible person must ensure that mine design is based on appropriate geotechnical information and takes into account geotechnical risks. A system must ensure that geotechnical design parameters are used to optimise the size, shape and orientation of mining excavations. Geotechnical design parameters established for each rock mass domain should form a base for assessment of rock mass stability. In underground mines, geotechnical considerations should be given to determining the maximum open span of excavations and the dimension and shape of pillars (with special emphasis on surface crown pillars). The effect of interaction of excavations and backfill should be considered. This could allow for development of a sequence of mining extraction and backfilling. Potential for mining-induced seismic activity should be considered.

12.4.6 Approval system

Ground control information or a geotechnical input into mine design should be prepared and documented by a competent person and must be approved by an authorised person. Conversely, a formal geotechnical approval process for the development and mining activities should be implemented. Variation from the original design should be documented and approved.

12.4.7 Feedback and follow-up

The ground control management system should establish procedures for verification of plans and design as a project progresses and rock mass behaviour changes. Geotechnical monitoring should provide further information and feedback for successful implementing findings into an evolving plan of remedial measures. The feedback might be used to modify the design or change some design parameters in the consecutive design phases.

12.5   QUALITY ASSURANCE IN GROUND CONTROL ACTIVITIES

Ground control activities include drilling, blasting, barring down and excavation mainte-nance. Regular and systematic equipment calibration checks should be made on the equip-ment used for drilling, blasting, support installation and geotechnical instrumentation.

12.5.1   Drilling

Drilling procedure and specification for ground control purposes, i.e. for blasting or sup-port installation, should be prepared and approved. The specification should include collaring position, direction of drilling and hole length. The direction of drilling and hole length should be provided with acceptable deviations. The depth of the holes should be monitored and cor-rected as required. Drilling conditions should be monitored, and a formal feedback should be provided to blasting or support engineers. That feedback, if required, should be used to modify design parameters. An example of an irregular and potentially unstable profile of an underground excavation due to poor quality of development drilling is given in Figure 12.2.

12.5.2   Blasting

Quality assurance in ground control management requires that blasting activities should be carried out according to existing procedures. Blasting procedures should specify type of explosives, blast hole parameters, charging method, a system of initiation and a system of reporting misfires. Blasting should be designed for each geotechnical domain and blasting process and resulting fragmentation should be monitored. Lack of quality assurance can

Figure 12.2 Irregular and potentially unstable profile of an underground excavation due to poor quality of drilling (Szwedzicki, 2003a).

result in instability of the rock mass, poor fragmentation, overblasting, uneven shape of excavations, uneven floor, toe remnants in open pits (Fig. 12.3) or remnant pillars in underground mines (Fig. 12.4). Large blasts have a potential to cause major damage to the rock mass surrounding excavations and could act as a catalyst provoking seismic events.

Figure 12.3 A toe remnant in a wall of an open pit (Szwedzicki, 2003a).

Figure 12.4 A pillar remnant in an open stope.

12.5.3   Maintenance of excavations

Maintenance of excavations includes barring down, ensuring compliance with design, support, water management and reporting of the rock mass instability. All these activities should be preceded by a risk analysis.

- *Barring Down*

Written procedure on barring down (scaling down) should specify the method of scaling rocks down and proper use of the equipment. For existing excavations, the procedure should specify the minimum intervals between barring down as rock mass can suffer from blasting and ground conditions may deteriorate with change in mining-induced stress or time. Figure 12.5 gives an example of an open pit wall where barring down was not performed.

- *Verification with Design*

A procedure must be developed to verify of the mine advance (production and development) with the design parameters. The procedure should specify acceptable horizontal and vertical deviations from the design. An unacceptable deviation from the design should be recorded and (in case of overbreaks, overhangs or leaving remnants) remedial action should be taken.

- *Ground Support*

Quality assurance in ground support (including rock support and reinforcement) should be executed in design, installation and performance monitoring.

Figure 12.5 An open pit wall with loose rocks (Szwedzicki, 2003a).

Ground Support Design: The design document should specify type of support and rein-forcement (e.g. length, diameter, steel type, grout consistency), support density and support layout (e.g. number of bolts in a row, spacing between the rows) and support specification (e.g. bolt hole position, inclination and depth, thickness of shotcrete, mechanical properties of support material, consistency of grout). Support design should take into account mechanical properties of the rock mass, structural features of the rock mass, *in situ* and mining-induced stress, and the effect of water on the stability of the rock mass and on corrosion of support elements. Areas that are rec-ommended for support should be indicated on mining plans. Figure 12.6 illustrates support installed without a formal geotechnical design.

Ground Support Installation: All support must be installed according to the design pattern and installation procedures. Procedures shall include details on storage and handling of ground support material, assessment of ground support stability, ground support installation and recording of installation data (Mines and Aggregate Safety, 1998). Figure 12.7 gives an example of poorly installed support.

Figure 12.6 Oversupporting of an underground excavation (Szwedzicki, 2003a).

Figure 12.7 An example of poor quality installation of support – a protruding rock bolt with unsecured mesh (Szwedzicki, 2003a).

12.5.4   Performance of ground support

A procedure of the quality control program to assess the performance of installed support should specify the parameters and the conditions of testing. Performance of ground support should be tested after installation and then monitored over the life of mining excavations. Support that is to be tested in a destructive way should be installed in addition to the support required for the specific pattern.

Testing during or after the installation should include:

- Testing support elements to ensure that they meet specifications, for example, consistency and properties of grout used for bolting or shotcrete mix. Using thin grout not only reduced the pull out strength but also could result in bolt slipping out of hole before curing Fig. 12 8.
- Testing for mechanical properties of installed support, for example, pull-out tests. A pull-out test procedure should specify the number of bolts to be tested and the method of recording, and it should provide minimum standards for mechanical parameters that must be achieved.

Long-term monitoring should include observation of the interaction between the rock mass and installed support and observation for corrosion of steel elements. All instances of rehabilitation of areas supported in the past should be investigated and feedback provided for support design.

- *Water Management*

Water management should include risk analysis of water flow into mine excavations. The risk analysis should consider diversion of storm water, ponding of surface water, surface water disappearance, water seepage into the excavations, formation of cracks on the surface

Figure 12.8 Thin grout resulted in the bolt slipping out of hole.

or surface subsidence, and erosion of embankments and dams. Quality assurance requires that a procedure(s) on water management should be prepared and implemented. The procedure should deal with the possible effect of water on stability of the rock mass. It should have provisions for drainage using dewatering holes, calculation of sump's volume, drainage diversions, protective bund walls, etc. Figure 12.9 shows how poor construction of a protection bund wall resulted in water flooding an open pit.

- *Instability of the Rock Mass*

A procedure should be prepared on reporting and investigating instability of the rock mass, i.e. falls of rock and support failure. Standard report forms should be available. Information should include location, failure dimensions and mode, comment on stress change, description of geotechnical features, excavation and rock support details and results of monitoring (Geotechnical Considerations, 1997). Figure 12.10 gives an example of a large rockfall in an underground mine.

Figure 12.9 Water flooding an open pit (Szwedzicki, 2003a).

Figure 12.10 A rockfall in an underground excavation (Szwedzicki, 2003a).

12.6 QUALITY ASSURANCE IN GEOTECHNICAL INSPECTION AND MONITORING

Geotechnical inspection and monitoring serve to locate any potential uncontrolled instability of ground before the ground becomes unstable and hazardous. Early detection of failure allows mine operators to plan and implement actions limiting the effects of impending failure. Geotechnical monitoring is carried out to assess changes in rock mass behaviour in time. It may include taking readings of geotechnical instrumentation and making periodical observations.

12.6.1 Inspection

A competent person should be designated to carry out a geotechnical inspection of all areas affected by mining operations. A geotechnical inspection should be undertaken, on a regular basis, to check whether working activities or a workplace comply with requirements written down in work procedures, specifications, practices or standards. All changes in ground conditions and geotechnical warning signs of impending instability should be recorded and reported. It is recommended that a geotechnical inspection checklist is used, and results of each inspection are written down in an inspection record book. Inspections should not be confined to geotechnical deficiencies only but could highlight positive results of geotechnical mitigation activities and proper support installation practice. Figure 12.11 gives an example of exposed by blasting a grouted bolt installed in accordance with installation procedure.

Figure 12.11 A properly installed grouted bolt exposed after blasting.

12.6.2   Instrumentation

Geotechnical instrumentation, as determined in the Ground Control Management Plan, must be effectively installed to fulfil monitoring objectives. Persons installing instrumentation should follow manufacturer specifications and readings should be taken on a regular basis by a competent person. For each piece of instrumentation, the value of early warning and alarm trigger must be determined. An authorised person should monitor the results and all employees should be trained in an alarm system and an emergency procedure.

12.6.3   Monitoring of rock mass performance

The monitoring program should be specified in the Ground Control Management Plan. The program should include type of monitoring (observations and data recording), its frequency and type for interpretation and should specify alarm trigger values for impending failure. During mining operations, a system of ground performance monitoring and reassessment of mine design should be undertaken (Geotechnical Considerations in Open Pit Mines, 1999). Performance of the rock mass should be closely monitored by competent persons. Geotechnical monitoring equipment to verify the stability though displacement, stress or seismic measurement can be installed. Results obtained through rock mass monitoring should serve to refine the mine design process. All substantial changes of the monitored values must be communicated to designated employees. Areas of potential instability should be delineated and made known to the working crews. A job safety analysis should be carried out for areas of high risk, and a permit system to work in such areas is required. Monitoring results and recommendations following from their review should be passed on to relevant authorised personnel at regular intervals.

Safety and productivity in mines can be improved by implementing a quality assurance program in ground control management. The program is achieved by the following:

*   defining of company policy and development of Ground Control Management System. The system should specify responsibilities and authority, document control system and competency and training required for each job.
*   implementation of quality assurance in geotechnical planning and design. It should cover data collection and analysis, preparation and execution of the Ground Control Management Plan, approval process and geotechnical feedback.
*   introduction of quality criteria in work procedures and practices on drilling, blasting, barring down, ground support, water management and reporting of instability of the rock mass.
*   specifying quality factors in geotechnical inspections and monitoring.

Audits on quality assurance in ground control can provide mine management with the information on status and potential improvement in geotechnical activities.

References

Amadei, B., Krantz, R.L., Scott, G.A. & Smealie, P.H. (1999) Description of a large catastrophic failure in southwestern Wyoming Trona Mine. *Proc. of the 37th US Rock Mechanics Symp.*

Analysis of coal refuse dam failure. (1973) Middle Fork, Buffalo Creek, Saunders, West Virginia. US Department of the Interior, Washington, DC.

Bailey, J. (2003) *Findings and Recommendations: Inquest into the deaths of R. Brodkin, M House, S Osman and C Lloyd-Jones on 24 November 1999 at the E26 Lift 1.* Northparkes Mine, Parkers, NSW.

Bandis, S.C. (1990) Scale effects in the strength and deformability of rocks and joints. *Proc. Conference on Scale Effects in Rock Masses.* A Balkema, Rotterdam. pp. 59–76.

Barber, J., Thomas, L. & Casten, T. (2000). Freeport Indonesia's Deep Ore Zone Mine. *Proc. AusIMM MassMin 2000 Conf.* Brisbane, 29 Oct–2 Nov.

Beus, M.J., Pariseau, W.G., Steward, B.M. & Iverson, S.R. (2001) Design of orepasses. In Hustrulid, W.A. & Bullock, R.L. (eds.) *Underground Mining Methods, Engineering Fundamentals and International Case Studies.* Society of Mining, Metallurgy and Exploration, Littleton, Colorado, pp. 627–634.

Bieniawski, Z.T., Franklin, A., Bernede, N.I.I., Duffaut, P., Rummel, F., Horibe, T., Broth, E., Rodrigues, E., Van Heerden, W.L., Vogler, U.W., Hansagi, I., Szlavin, J., Brady, B.T., Deere, D.L.7, Hawkes, Z.T & Milovanovic, D. (1979) Suggested methods for determining the uniaxial compressive strength and deformability of rock materials. *International Journal of Rock Mechanics and Mining Sciences and Geomechanics*, 16(2), 135–140.

Biggam, F.B., Robinson, B. & Ham, B. (1980) Outbursts at Collinsville: A case study. *Proc. AusIMM Southern Queensland Branch, the Occurrence and Control of Outbursts in Coal Mines Symposium*, September.

Brady, B.H.G. & Brown, E.T. (1993) *Rock Mechanics for Underground Mining.* Chapman and Hall, London.

Broom, M.T. & Sandy, M.P. (1988) Rock mechanics investigations at Mufulira mine, Zambia. *Transactions of the Institution of Mining and Metallurgy. Section A. Mining Industry*, Vol 97, January 5(3), 99–103.

Brown, T. (2007) *Block Caving.* 2nd edition. Julius Kruttschnitt Mineral Research Centre (JKMRC), Brisbane.

Bryan, A., Bryan, J.G. & Fouche, J. (1964) Some problems of strata control and support in pillar workings. *Mining Engineering*, 123.

Bull, G. & Page, C.H. (2000) Sublevel caving: Today's dependable low-cost "ore factory". *Proc. MassMin Conference*, Brisbane, November.

Call, R.D. (1992) Slope stability. In *SME Mining Engineering Handbook.* Vol 1. Society for Mining, Metallurgy and Exploration, Inc., Littleton, CO.

Carter, T.G. Miller, R.I. (1995) Crown-pillar risk assessment: Planning aid for cost-effective mine closure remediation. *Transactions of the Institution of Mining and Metallurgy. Section A. Mining Industry*, 104, Al–78.

Dobry, R. & Alvarez, L. (1967) Seismic Failure of Chilean Tailings Dams. *Journal of Soil Mechanics and Foundation Division*, 93(6), 237–260.

Dunn, P.G., Whitmore, J., Szwedzicki, T., Robb, A., McHugh, C., Maciejewski, L. & Blyth, M.G. (2006) Hydroscaling for Rapid Drift Development. *Proc. 2nd Int. Symp. on Rapid Mine Development, June*. Aachen, 7–8, pp. 41–55.

Emmerich, S.H. (1992) Report on rock passes in Anglo American Corporation Gold Division Mines. *Proc. Symp. on 'Orepass and combustible material underground'*. Association of Mine Managers of South Africa, pp. 83–111.

Environmental and Safety Incidents concerning tailings dams in mines. (1996) *Results of Survey by Mining Journal Research Services*. A report prepared for United Nations Environment Programme, p. 51.

Fairhurst, C. & Cook, N.G.W. (1966) The phenomenon of rock splitting parallel to the direction of maximum compression in the neighbourhood of a surface. *Proc. 1st Congress ISRM*, Vol I. International Society for Rock Mechanics, Lisbon. pp. 687–690.

Farmer, I.W. & Kemeny, J. M. (1992) Deficiencies in rock test data. *Proc. Int. Symp. Eurock '92*, London. pp 298–303.

Fox, M.J. (1995) *Quality Assurance Management*. Chapman & Hall, London.

Galvin, J.M. (1998) Lassing mud inrush disaster, Austria, July 1998. NSW Coal Mine Managers Association, Emergency Preparedness Seminar.

Gardener, R.D. & Pincus, H.T. (1968) Fluorescent dye penetrants applied to rock fractures. *International Journal of Rock Mechanics and Mineral Sciences*, 5, 155–158.

Geotechnical Considerations in Open Pit Mines. (1999) Department of Minerals and Energy, Perth, Western Australia.

Geotechnical Considerations in Underground Mines. (1997) Department of Minerals and Energy, Perth, Western Australia.

Gibowicz, S.J. & Kijko, A. (1994) *An Introduction to Mining Seismology*. Academic Press, San Diego.

Glazer, S.N. (2016) *Mine Seismology: Seismic Warning Concept Case Study from Vaal Reefs Gold Mine*. Springer International Publishing, South Africa.

Goel, S.C., Page, S.H., (1982) An empirical method for prediction the probability of chimney cave occurrence over mining area. *International Journal of Rock Mechanics and Mining Science and Geomechanics*, Abstract, 19, 325–337.

Goodspeed, T., Skinner, J. & Friend, R.M. (1995) Accident Investigation Report Underground Non-Metal Mine Fatal Collapse of Mine Workings, February 3, US Dept. of Labour Mine Safety and Health Administration.

Goodwill, D. J., Craig, D. A. & Cabrejos, F. (1999) Orepass design for reliable flow. *Bulk Solids Handling*. 19, 13–21.

Gramberg, J. (1989) *A Non-Conventional View on Rock Mechanics*. AA Balkema, London.

Hadjigeorgiou, J., Lessard, J.F. & Mercier-Langevin, F. (2004). Issues in selection and design of orepass support. *Proc. 5th International Symposium on Ground Support in Mining and Underground Construction*. Australian Centre for Geomechanics, Perth.

Hadjigeorgiou, J., Lessard, J.F. & Mercier-Langevin, F. (2005) Orepass practice in Canadian Mines. *Journal of the Southern African Institute of Mining. Metallurgy*, December 105, 809–816.

Hambley, D.F. (1987) Design of orepass systems for underground mines. *Canadian Institute of Mining, Metallurgy and Petroleum*, January, 80(897), 25–30.

Harding, R. (2000) The next environmental challenge: Moving from risk management to the precautionary principle. *Proc. AusIMM Conference after 2000: The Future of Mining*, Sydney, NSW. 10–12 April.

Hathaway, A.W. (1968) Subsidence at San Manuel Copper Mine, Pinal County, Arizona. In *Engineering Geology Case Histories*. Number 6, The Geological Society of America. Boulder, CO.

Hebblewhite, B. (2003) Northparkes: The. findings and the future. 7th Kenneth Finlay Memorial lecture, School of Mining Engineering, UNSW.

Hoek, E. & Brown, E.T. (1980). *Underground Excavations in Rock*. Institution of Mining and Metallurgy, London.

Hollins, B. & Tucker, J. (2004) Drawpoint analysis using a marker trial at the perseverance nickel mine, Leinster, Western Australia. *Proc. MassMin 2004*. Santiago, Chile.

Horii, H. & Nemat-Nasser, S. (1985) Compression-induced microcrack growth in Brittle solids: Axial splitting and shear failure. *J. of Geophysical Research*, March, 90(B4), 3105–3125.

Hustrulid, W. (1999) *Mining Method*. SME, Littleton, Colorado.

Hustrulid, W. & Kvapil, R. (2008) Sublevel caving: Past and future. *Proc. of the 5th Conference on Mass Mining, 9–11 June* 2008. Lulea, Sweden.

Hudson, J.A. & Harrison, J.P. (1997) *Engineering Rock Mechanics*. Oxford, Pergamon.

Improving Ground Stability and Mine Rescue. 1986. The Report of the Provincial Inquiry into Ground Control and Emergency Preparedness in Ontario Mines, Ontario Government.

International Standard ISO 9001, Quality systems: Model for quality assurance in design, development, production, installation and servicing.

International Standard ISO 9002, Quality systems: Model for quality assurance in design, development, production, installation and servicing.

International Standard ISO 9003, Quality systems: Model for quality assurance in final inspection and test.

Jarosz, A. & Langdon, J. (2007) Development of inspection and surveying tool for vertical mining openings and shafts. *Proc. FIG Working Week 2007*. International Federation of Surveyors, Denmark.

Jumikis, A.R. (1983) *Rock Mechanics*, 2nd Edition. Trans Tech Publications, Clausthal Zellerfeld.

Kaiser, P.K., Diederichs, M.S., Martin, C.D., Sharp, J. & Steiner, W. (2000) Underground works in hard rock tunnelling and mining. *GeoEng 2000. Int. conf. on Geotechnical and Geological Eng.* Melbourne.

Kilpatrick B., Szwedzicki T., (1994) Model Studies of Mining Pillars Stability. *Proc. AusIMM Conference*, Brisbane. 28–29 April, pp. 6568.

Klokow, J.W. (1992) The collapse of room-and-pillar workings and planned pillar extraction at Otjihase mine, Namibia. In *MASSMIN 92*. SAIMM, Johannesburg.

Koivu, G.E. (1982) Rapport final sur les circonstances, les conditions prealables et les causes de la tragedie du 20 May, 1980. Ontario Ministry of Labour, May.

Kurzeja, A. (1992) Empirical criterion for rock bolting requirements. *Proc. Western Australian Conf. on Mining Geomechanics, Kalgoorlie*, June.

Laubscher, D. (1995) Cave mining: State-of-the-art. *Proc. AusIMM Underground Operator's Conference, 13–14 November*. Kalgoorlie, Australia.

Lessard J.F. & Hadjigeorgiou, J. (2003a) Design tools to minimize the occurrence of orepass interlocking hang-ups in metal mines. *Proc. ISRM Conf. on 'Technology roadmap for rock mechanics', September*, Santon, South Africa, South African Institute of Mining and Metallurgy, pp. 1–6.

Lessard, J.F. & Hadjigeorgiou, J. (2003b) Orepass systems in Quebec Underground Mines. *Proc. Int. Symp. on 'Mine planning and equipment selection', April*. Kalgoorlie, Australia, Australasian Institute of Mining and Metallurgy (AusIMM), pp. 509–521.

Madden, B.J. (1989) Squat pillar design in South African Collieries. *Proc. Symp. On Advances in rock mechanics in underground coal mining. International Society for Rock Mechanics*, Witbank, RSA, September, pp 83–90.

McCarthy, P.L. & Askew, J.E. (1986). Pilot Study on Waste and Orepass Design in Underground Mines, CSIRO Division of Geomechanics, March, Melbourne.

Mendes, F.M. and Da Gama, C.D, (1972) Laboratory Simulation of mine pillar behaviour. *Proc. 14th U.S. Symp. on Rock Mechanics*. University Park, Pennsylvania.

Mines and Aggregate Safety and Health Association. Guidelines for Quality Control of Ground Support in Underground Mines, Canada, 1998.

Moerdyk, C.M. (1965) *Historical, Technical Review of Coalbrook Disaster, 1960*. The Government Mining Engineer, Johannesburg.

Morgenstern, N.R., Vick, S.G., Viotti, C.B., Bryan, D. & Watts, B.D. (2016) Report on the Immediate Causes of the Failure of the Fundão Dam. August, p. 79.

Mufulira Mine Disaster – Final Report of the Commission of Inquiry, Lusaka, 1971.

Ormonde R., Szwedzicki T., (1993) Monitoring of Post-Failure Pillar Behaviour - Laboratory Tests. *Proc. Australian Conference on Geotechnical Instrumentation in Open Pit and Underground Mining,* Kalgoorlie, 21–23 June, pp. 393–399.

Paul, B. & Gangal, M. (1966). Initial and subsequent fracture curves for biaxial compression of brittle materials. In *Proc. 8th U.S. Symposium on Rock Mechanics,* Baltimore, pp. 131–141.

Peng, S. & Johnson, A.M. (1972) Crack propagation and faulting in cylindrical specimens of Chelmsford Granite. *International Journal of Rock Mechanics and Mining Sciences,* 9, pp. 37–86.

Piatek, M. (1980) Wyrzut gazow i skal w kopalni Nowa Ruda. *Bezpieczenstwo Pracy w Gornictwie,* (2), 4–9, (in Polish).

Rachmad, L. & Sulaeman, A. (2002) Cave management practices at PT Freeport Indonesia's block caving mine. *Proc. NARMS-TAC,* Hammah, R. *et al.* (eds.), Balkema.

Rachmad, L. & Widijanto, E. (2003) Application of convergence monitoring at PT Freeport Indonesia Deep Ore Zone mine. *Proc. NARMS-TAC,* Hammah, R. *et al.* (eds.), pp. 181–189.

Reinhart, J.S. (1966) Fracture of rocks. *International Journal of Fracture Mechanics,* 2, 534–590.

Report of the Tribunal Appointed to Inquire into the Disaster at Aberfan on October 21st, 1966, Her Majesty Stationary Office, London, 1967.

Ross, I. & van As, A. (2005) Northparkes mine – design, Sudden failure, air-blast and hazard management at the E26 block cave. *Proc AusIMM Ninth Underground Operators Conf,* Perth, pp. 7–18.

Rustan, A. (2000) Gravity flow of broken rock – what is known and unknown. *Proc. MassMin 2000,* Brisbane, Australia.

Sahupala, H.A. & Srikant, A. (2007) Assessment of pillar damage at the extraction level in the Deep Ore Zone (DOZ) mine. *Pt Freeport Indonesia, First International Symposium on Block and Sub-Level Caving,* SAIMM, Cape Town.

Sahupala, A.H. & Szwedzicki T. (2004) Geotechnical events leading to closure of an underground crusher chamber. *Proc Int Conf MassMin,* Santiago. pp. 263–268.

Sahupala, H.A., Szwedzicki, T. & Prasetyo, R. (2010) Diameter of a draw zone – a case study from a block caving mine, Deep Ore Zone, PT Freeport Indonesia. *Proc 2nd Int. Seminar on Block and Sublevel Caving,* ACG, Perth, April.

Sandy, J.D., Piesold, D.D.A., Fleisher, V.D. & Forbes, P.J. (1976) *Failure and Subsequent Stabilization of No. 3 Dump at Mufulira Mine, Zambia. Transactions of the Institution of Mining and Metallurgy,* October, pp. A144–A162.

Sheehy, J.A. (1956) *Report of the Royal Commission Appointed to Inquire into Certain Matters Relating to the State Coal Mine, Collinsville.* Government Printer, Brisbane.

Spottiswood, S.M. (2010) Mine seismicity: Prediction or forecasting. *Journal of the Southern African Institute of Mining and Metallurgy,* 110, 11–20.

Stacey, T.R. (2004) General guidelines for the design of rock passes. Proc. SAIMM Colloquium on Design, Development and Operation of Rockpasses. *Journal of the Southern African Institute of Mining and Metallurgy,* Marshalltown.

Stacey, T.R. & Erasmus, B.J. (2005) Setting the scene: Rockpass accident statistics and general guidelines for the design of rockpasses. *Journal of the Southern African Institute of Mining and Metallurgy,* 105, December, 745–752.

Stacey, T.R. & Page, C.H. (1986). *Practical Handbook for Underground Rock Mechanics.* Trans Tech Publications, Clausthal Zellerfeld.

Starfield, A.M. and Wawersik, (1968) W.R. Pillars as structural components in room-and-pillar mine design. *Proc.10th Symposium on Rock Mechanics,* Austin, Texas, 20–22 May.

Staunton, J.H. (1998) Report of a Formal Investigation under Section 98 of the Coal Mines Regulation Act, 1982, June.

Stewart, B., Iverson, S. & Beus, M. (1999) Safety considerations for transport of ore and waste in underground orepasses, *Mining Engineering.* March, 53–60.

Sullivan, T.D. (1993) Understanding pit slope movement. In Szwedzicki, T. (ed.) *Proceedings of the Conference on Geotechnical Instrumentation and Monitoring in Open Pit and Underground Mining*, Balkema.

Szwedzicki, T. (1989a) Geotechnical assessment deficiency in underground mining. *Mining Science and Technology*, (9), 23–37, Elsevier Science Publisher, Amsterdam.

Szwedzicki, T. (1989b) Pillar recovery in Mhangura copper mines, Zimbabwe. *Transactions of the Institution of Mining and Metallurgy. Section A. Mining Industry*, (98), September–December, A127–A136.

Szwedzicki, T. (1992) Geotechnical problems caused by abandoned underground mines in the Eastern Goldfields. *Proc. Western Australian Conference on Mining Geomechanics*, Kalgoorlie, 8–10 June, pp. 461–468.

Szwedzicki, T. (1999a) Pre- and post-failure behaviour of surface crown pillars – case studies. *International Journal of Rock Mechanics and Mining, Sciences*, (36), 351–359.

Szwedzicki, T. (1999b) Sinkhole formation over mining areas and risk management implications. *Transactions of the Institution of Mining and Metallurgy*, 108, London, January–April, A27–A36.

Szwedzicki, T. (2000) The effect of mining geometry on stability of the rock mass around underground excavations. *Mineral Resources Engineering Journal, Imperial College Press*, 9(2), 17.

Szwedzicki, T. (2001a) Geotechnical precursors to large-scale ground collapse in mines. *International Journal of Rock Mechanics and Mining Sciences*, 9, London.

Szwedzicki, T. (2001b) Program for mine closure. *Mineral Resource Engineering*, 10(3), 347–364.

Szwedzicki, T. (2003a) Quality assurance in mine ground control management. *International Journal of Rock Mechanics and Mining Sciences*, 40, 565–572.

Szwedzicki, T. (2003b) Rock mass behavior prior to failure. *International Journal of Rock Mechanics and Mining Sciences*, (40), 573–584, London.

Szwedzicki, T. (2004) Warning signs to geotechnical failure of mining structures. *International Journal of Surface Mining, Reclamation and Environment*, 18(2), 150–163.

Szwedzicki T. (2005) Reviewing support requirements for existing excavations in underground mines. *Transactions of the Institution of Mining and Metallurgy*, 114, March, pp. A21–28.

Szwedzicki, T. (2007a) Formation and removal of hang-ups in orepasses. *Mining Technology, Transactions of the Institution of Mining and Metallurgy. Section A. Mining Industry*, 116(3), 139–145.

Szwedzicki, T. (2007b) A hypothesis on modes of failure of rock samples tested in uniaxial compression. *Rock Mechanics and Rock Engineering*, 40(1), 97–104.

Szwedzicki, T. & Cooper, R. (2007) Ore flow and fragmentation at perseverance mine. *Proc. 1st Int. Symposium on Block and Sub-Level Caving*, Cape Town, South Africa, 8–11 October, pp. 135–146.

Szwedzicki, T. & Shamu, W. (1996) Detection of planes of weakness in rock samples using non-destructive testing method. *Proc. Int. Symp. on Mining Science and Technology, '96*. China, AA Balkema, 16–18 October, pp. 759–763.

Szwedzicki, T. & Shamu, W. (1999) The effect of material discontinuities on strength of rock samples. *Proc. Australasian Institute of Mining and Metallurgy*, 304(1), 23–28.

Szwedzicki, T., Valent, M.C. & Gaudreau, D. (2007) Rock mass response to mining at Perseverance mine, Western Australia. *Proc. 33rd Int. Symp. APCOM*, Santiago, pp. 199–205.

Szwedzicki, T., Widijanto, E. & Sinaga, F. (2004) Propagation of a caving zone, a case study from PTFI Freeport, Indonesia. *Proc. Int. Conf. MassMin*, Santiago. pp. 508–512.

Tailings Dam Risk of Dangerous Occurrences, Lessons Learnt from Practical Experiences. (2001) Commission Internationale des Grandes Barrages, Paris, p. 144.

Thompson, P.W. & Cierlitza, S. (1993) Identification of a slope failure over a year before final collapse using multiple monitoring methods. *Proc. Conf. on Geotechnical Instrumentation and Monitoring in Open Pit and Underground Mining*, Szwedzicki (ed.), Balkema.

Tierney, S.R. & Morkel, I.G. (2017) The optimization and comparison of re-entry assessment methodologies for use in seismically active mines. In Wesseloo, J. (ed.) *Eighth Int. Conf. on Deep and High Stress Mining*. Deep Mining, Perth, Australia.

Tyler, D., Campbell, A. & Haywood, S. (2004) Development and measurement of subsidence zone associated with SLC operations at perseverance – WMC, Leinster Nickel operations. *Proc. Mas-Min*, Santiago.

Varden, R. (2001) *East Wall Subsidence at Sons of Gwalia, in Advance Rock Mechanics Practice for Underground Mines*. Australian Centre for Geomechanics, Perth, Western Australia.

Villafuerte, G., Gumo, S. & Rodriges, R. (2007) Mitigating geothermal hazards at Lihir gold mine. *AusIMM Bulletin*, November.

Vutukuri, V.S., Lama, R.D. & Saluja, S.S. (1974) *Handbook on Mechanical Properties of Rock*. Vol. 1. Trans Tech Publications, Clausthal Zellerfeld, p. 269.

Wagener, F., Craig, H.J., Blight, G., McPhail, G., Williams, A.A.B. & Strydom, J.H. (1998) The Marriespruit tailings dam failure – a review. *Proc. of the 5th Int. Conf. on Tailings and Mine Waste '98. AA Balkema, Rotterdam*.

Wagner, H. (1984) Fifteen years' experience with design of coal pillars in shallow South African Collieries: Evaluation of the performance of the design procedure and recent improvements. *Proc. Int. Society of Rock Mechanics Symp. Design and Performance of Underground Excavations*, Cambridge, UK.

Wood, P., Jenkins, P.A. & Jones, I.O. (2000) *Sublevel cave drop down strategy at Perseverance Mine, Leinster Nickel Operations*. Mass Min 2000, Brisbane, November.

Zipf, R.K. & Swanson, P. (1999) Description of a large catastrophic failure in southwestern Wyoming Trona Mine. In Amadei, B., Kranz, R.L., Scott, G.A., & Smeallie, P.H. (eds.) *Proc. of the 37th U.S. Rock Mechanics Symposium*. Vail, Colorado.

Index

Aberfan 95–96, 130

Balmoral mine 94
borehole breakout 116
brittle behaviour 33
Bronzewing Gold Mine 93
bulging 116, 130

caving 57–64, 160, 164–167, 172
    zone 60
Chaffers shaft 41
Coalbrook Colliery 65–66, 132, 144
collapse *see* ground, collapse
compaction 163, 169, 188, 190
Coronation mine 40

dilution 176, 180–182
discontinuities 103, 105, 108
    microscopic 103–110
draw 163, 165, 171, 174–176, 190
    density 22, 34–35
    management practices 180–183
    rate 171
    zone diameter 166–167, 169–170, 183–184
draw point 135–136, 138–140, 165
ductile deformation 147–149

extraction
    partial 148
    ratio 25, 30–32, 64, 138, 148–149
    selective 148
    total 27, 30, 135

failure
    brunt 122, 150
    duration 66, 129–132, 152
    onset 126, 129

post-failure 19, 31, 33, 35, 143–144, 147–149
pre-failure 32, 102, 111, 159
rock 29, 149
rock mass 150–151
fall of ground 20, 68, 75, 94
floor heave 150, 157, 172
fracturing 20, 23–24, 31–32, 34–35, 58, 61–63,
    104–105, 138
fragmentation 26, 32, 163–164, 171, 179, 208

Gretley colliery 90–91, 130
ground
    collapse 17, 57, 130, 147, 158
    control 195–200, 202–206
    control management 196, 198–200, 205,
        207, 214
    deterioration 25, 68, 73–76, 79, 163

hang-up 71, 140, 173, 182, 185–189, 191
hydroscaling 28

indicators 50, 52, 99, 111–114
infrastructure 29, 68; *see also* mining methods
inrush 146
    backfill 93
    mud 89, 94
    tailings 38, 91
    water 37, 90
inundation 17, 22, 37, 89, 130
Iron King mine 42

Lihir gold mine 56

Marriespruit 96
Merthyr Vale colliery *see* Aberfan
mining
    geometry 25, 30–32, 148, 154

mining methods 27
   block caving 57, 60, 63, 138, 165
   room-and-pillar 32, 57, 67, 154
   shrinkage 25, 40–41, 94
   sublevel caving 137, 172, 183
mode of failure 24–25, 108, 110, 152–153, 157
   matrix 152–153
   multiple extension 104, 155
   multiple fracturing 104–105, 110, 153
   multiple shear 104–105, 153
   simple extension 104
   simple shear 104–105, 129, 153
   stress induced 19, 154
   structurally controlled 19, 24, 154
monitoring 159, 163, 195, 198, 212–214
   geotechnical 159, 205–206
   rock mass performance 197
Mufulira mine 91–92

No.1 State mine 155
Nobles Nob mine 43, 145
non-destructive testing 106, 109, 151
Northparkes mine 132, 141
Nowa Ruda mine 53, 132

ore 63–64
   draw 60, 140, 155, 167, 167, 180
   flow 174, 178, 183
   fragmentation 163, 171–173, 180, 185
   recovery 174–176, 180
orepass 69–73, 185, 188, 191
   hang-ups 71, 173, 185–188, 191
Otjihase mine 64, 132
outburst
   gas 22, 38, 51, 53–55
   geothermal 56, 129

Perseverance shaft 44, 146, 173, 177
Pillar 29–31, 64, 104, 149, 155
   collapse 35, 39, 65, 67, 140
   model 31–35
   regular 34–36
   remnant 27, 135, 208
   slender 31–36
   squat 34–36
   surface crown 38, 48
   width;height ratio 25, 32, 147
   yielding 32, 115, 133
precursors 21, 111, 114, 120–123, 125
prediction 23, 38, 51, 124
Prince of Wales mine 45

PT Freeport Indonesia 60, 165
progressive collapse 39, 67, 91, 132

quality 195, 197
   assessment 197, 199
   assurance 195–197, 204, 213
   audit 200
   blasting 208
   control 197
   drilling 202
   grade 197
   ground control 196, 202
   level 199
   system 196

re-entry time 143
risk management 122–123, 195, 200
rockbursts 37, 150
rock mass 23
   behaviour 23, 110, 129, 143, 159
   post-failure 21, 33, 50, 143–147
   pre-failure 31, 111
   progressive 126, 133, 150
   regressive 78, 126
   transgressive 78, 126, 134, 150
roof sagging 115

San Manuel mine 56
scale 74, 79
   excavation 118
   local 118
   mine 77, 111, 117
Scotia mine 45, 145
seismic activity 20, 24, 41, 51–52, 119, 159
sinkhole 38–40, 49, 51, 59, 145
Solvey Minerals mine 144
Sons of Gwalia mine 83
Spalling 24, 29, 121, 149
stress
   abutment 133–135, 137
   compliance matrix 151–152
   in-situ 20, 24–25
subsidence 27, 30, 122, 144
   continuous 38
   discontinuous 37–39, 57–59

tailings 37, 83, 89–92
   dams 19, 21, 96, 100
   instability 100
Telfer Gold Mine 87, 146
time domain reflectometry 60

tramp material 166–168, 183
triggers 119, 127
   external 119
   internal 119

underground infrastructure 29
   crusher 68, 73–75
   orepass 29, 68–72, 163, 185, 191
   shaft 32, 41, 44, 81

uniaxial compressive strength 86, 105, 108, 153
Union Reefs mine 103
unravelling 75, 115
unsupported span 30, 39, 159

warning sign 21, 74, 79, 111, 214
Warrego mine 47, 130, 147
waste rock 37, 95, 163, 181, 185
   dumps 83, 90

Geomechanics Research Series
Series Editor: Marek A. Kwaśniewski
Publisher: CRC Press/Balkema, Taylor & Francis Group

1 **Fractals in Rock Mechanics**
   Editor: Heping Xie
   1992
   ISBN: 978-90-5410-133-8

2 **Mechanical Behaviour of Rocks Under High Pressure Conditions**
   Editor: Mitsuhiko Shimada
   1999
   ISBN: 978-90-5809-316-5

3 **Experimental Rock Mechanics**
   Editor: Kiyoo Mogi
   2006
   ISBN: 978-0-415-39443-7

4 **True Triaxial Testing of Rocks**
   Editors: Marek Kwaśniewski, Xiaochun Li & Manabu Takahashi
   2011
   ISBN: 978-0-415-68723-2

Geomechanics Research
Series Editor: Tsuyoshi Ishida
Publisher: CRC Press/Balkema, Taylor & Francis Group

Rock Mass Response to Mining Activities
Author: Tadeusz Szwedzicki
2018
ISBN: 978-1-138-08292-2

Printed in the United States
by Baker & Taylor Publisher Services